The Hidden Biosphere

The Hidden Biosphere

Exploring life at the edge of perception

by

Adrian R Carr

Hidden Biosphere Press

First Edition

ISBN 978-1-7645173-0-0 (Paperback)
ISBN 978-1-7645173-1-7 (Hardback)

Published by Hidden Biosphere Press

Cover design, illustrations, and editing
by Adrian R Carr

Printed in Australia

TABLE OF CONTENTS

List of Figures

Preface

Had I captured on video what I initially hoped to, this book would likely have taken a quite different form. After more than forty years of waiting for official disclosure on the UFO phenomenon, I made the decision to begin my own investigation.

The aim of this project was simple: find evidence of an alien presence in our skies, physical craft that were from "elsewhere", not Earth. Beings who travelled in vehicles which traverse the heavens with the kind of advanced technological capabilities described in the many books and articles I had read about over preceding years. It was time to take an active approach to this enigmatic subject.

The visible light spectrum for humans is usually quoted as being 380 nm – 750 nm (nanometres), and near-infrared (NIR) radiation is generally defined as occupying the wavelength range from approximately 700 nm – 2500 nm, although exact boundaries vary slightly by discipline.

Therefore, I would require a camera which could capture these elusive craft if they appeared outside of the visible light spectrum. This would be in addition to the camera I already owned, a Canon EOS 400D and several lenses, and essential for my investigation.

There are many inexpensive models available, but most require a separate source of infrared illumination. My requirement was for a passive infrared camera, one that was capable of recording in infrared or near infrared range, at distance.

I spent several months researching suitable options, seeking a balance between cost and functionality. Eventually I settled on the Sionyx Aurora Pro. This is a full-spectrum camera which records in the 400 nm – 1100 nm wavelength range. It is not perfect, but has been a reliable piece of equipment, easy to use and quick to set up.

I had hoped that this research would result in the documentation of structured craft. Perhaps squadrons of flying saucers, or triangular craft like the legendary TR-3B; unknown or unacknowledged, solid, manufactured vehicles.

However, what I captured was, in hindsight, far stranger in many ways. Not conventional aircraft, nor even exotic spacecraft, but spheres of light, geometric forms, and luminous amorphous entities. Diverse, intelligent, plasma-like beings often appearing momentarily then fading back into the darkness.

Some were observed morphing, high in the atmosphere, whilst others closer to the ground moved with such acceleration and speed as to be almost impossible to see without modern, high framerate recording devices. Taken as a whole, this subject seemed to encompass a multi-layered ecosystem in an unexplored electromagnetic ocean extending to the heavens, virtually undetected and mostly invisible to the naked eye.

At first, I dismissed this as "operator error", atmospheric phenomena, tricks of the light, or camera artefacts, as I had no recollection of reading any literature describing such an ecology.

Repeated observations, corroborated with multiple devices and settings, convinced me these were not aberrations, lens flares or insects, and the images were proof that I hadn't imagined or hallucinated these objects.

I began to film night after night and often during the daylight hours. Scanning the footage and images during any spare moment. Frame-by-frame examination of luminous orbs and anomalies revealed internal structuring.

As the analysis progressed, it became possible to expose the faint, structural morphology surrounding the luminous core object. A body, although the term, "body", is not an accurate description, as these appeared to be beings of energy and resonant geometries; nevertheless, their subtle organisation often mirrored aspects associated with biological life.

Their movement was suggestive of life rather than randomness and the journey that followed took me deeper than I had anticipated, not only into the physics of plasmas, but into questions of consciousness, mythology, and cosmology.

As I persisted in exploring this subject, I realised that my location in the Pilbara region of Western Australia was fundamental to the subsequent development of the Plasma Entity Hypothesis (PEH) and particularly suited to the observation of such phenomena. This hypothesis began to form in the first few months of 2024, and left me feeling ostracised, frustrated and a little uneasy.

Later that year I came across a book by Trevor James Constable called, "*The Cosmic Pulse of Life*", which despite its controversial reception when originally published in 1976, echoed many of my own thoughts and findings.

Reading this work, proved unexpectedly reaffirming. With renewed enthusiasm, and a desire to bring this knowledge into the mainstream public focus, I spent every spare moment researching aspects of this phenomenon.

I became adept at spotting the fleeting, black or sometimes white objects, flitting across the screen, not just of my recordings, but also videos of others, many of which were completely unrelated to the UAP subject.

These fleeting objects move with apparent high velocity, almost always unnoticed, dismissed, or disregarded by the ~~origional~~original videographer or editor. I would try to isolate a single frame and attempt to expose more detail. However, these attempts were not always ~~successfull~~successful due to the reduced resolution of the uploaded recording.

In late 2024, when the internet lit up with "drone" sightings over US military bases and the US eastern seaboard, I got plenty of practice. During this time, I was sure that these objects would be officially identified for what they were and a whole new area of research would open up.

Promises by incoming president Trump, to get to the bottom of the drone incursions once in office, fell short. The official explanations that were offered seemed implausible, like a desperate attempt to quell growing public fear and frustration and the subject was soon dropped by the mainstream media. As I watched this narrative play out in real time, I felt obligated to write this book.

What began as a search for extraterrestrial spacecraft has instead evolved into the development of a theoretical framework concerning a possible hidden biosphere, in which intelligent plasma-based lifeforms may exist largely unseen, interacting with Earth's atmosphere, electromagnetic fields, and possibly at a more intimate level with humanity.

Although some of the ideas proposed here may appear speculative, they are consistently examined in relation to established scientific principles. In this context, the aim of this work is to present a set of observations and hypotheses that may assist in re-evaluating certain anomalous phenomena reported in Earth's skies.

Some readers may find these propositions uncomfortable or at odds with existing worldviews, while others may struggle to reconcile them with religious or metaphysical frameworks. This tension may represent one of the more challenging aspects of the work. Nevertheless, this book is not presented as a statement of final truth.

Rather, it offers an introduction and exploratory synthesis of mythological, historical, and contemporary accounts of anomalous phenomena across geographically and culturally unconnected societies. It seeks to outline plausible mechanisms by which such accounts may have arisen, while offering interpretive models that may also illuminate aspects of modern UAP reports.

We will discover almost identical reports documented in ancient texts, and descriptions from some cultures, passed down verbally through countless generations. These I propose, are part of the same multi-faceted phenomenon.

I hope to bring together my own research, experiences, and some contemporary accounts, in an attempt to reconcile with reflections on these historical, religious, and mythological accounts that I believe hint at the presence of such beings throughout human history.

The chapters that follow begin with my own observations~~ and expand,~~ expanding into complex dusty plasma theory, interdimensional hypotheses, the influence of geology on atmospheric phenomena, and ultimately into cosmology and metaphysics.

My aim is to show that the plasma entity hypothesis (PEH) not only accounts for my own ~~recordings~~observations and experiences but also connects to broader mysteries, many UFO encounters, anomalous lights, ancient myths of sky beings, and the deepest questions about life in our universe.

I believe this book is just scratching the surface of a subject which is without limitations in what could be revealed for mankind. If I had indeed found spacecraft and aliens, this book would have been far simpler to write.

Instead, I have stumbled across something both stranger, and possibly more profound. A parallel ecology of plasma-based life; luminous, intelligent, ephemeral beings that share our skies and our story in deeper ways than we ever imagined.

Introduction

Physical matter, also known as baryonic matter, exists in four states: solid, liquid, gas, and plasma. Plasma, the fourth state, is a partially or fully ionised gas, and is by far the most abundant form of ordinary matter in the universe. If we exclude dark matter and dark energy, around ninety-nine per cent of the visible universe is plasma, with the remaining one per cent made up of solids, liquids, and gases.

Our perception, shaped by our senses and processed by our brains, has been tuned by evolution to help our species survive long enough to reproduce and pass on our DNA to the next generation. We can detect other biological lifeforms whose physical mechanisms we recognise as similar to our own, which we classify as predator, prey, or useful to our existence.

This perception is profoundly limited, as we see only a narrow band of visible light, a tiny sliver of the electromagnetic spectrum, and hear but a fraction of the range of sound frequencies. The human sensory system inputs around eleven million bits of information to our brains every second, but the brain filters out approximately 99.9995 per cent of this information so that we do not become overwhelmed.

All that gets through to our conscious mind, totals just forty to fifty bits per second, just the essential information that offers the best possible chance of surviving long enough to pass on our DNA to future generations. In short, we miss a lot.

We often regard our species as the pinnacle of intelligence on Earth. Yet we live on a relatively insignificant speck of dust, orbiting an unremarkable star in a comparatively average size galaxy that, according to NASA, contains an estimated one hundred to four hundred billion other stars.

That unimaginable figure represents just the stars in our own galaxy. Across the observable universe, there are an estimated two trillion other galaxies. Some of these galaxies are dwarf systems with only a few billion stars, others are truly immense, containing trillions, and beyond the observable boundary of our universe there may be far more.

Given the abundance and diversity of biological life on Earth, it would be profoundly naïve to believe there is no other biological intelligence elsewhere in the cosmos. When we remember that solids, liquids, and gases make up only one per cent of baryonic matter, and plasma makes up the other ninety-nine per cent, another question arises. Could intelligent life also exist within the plasma state?

This book proceeds from observation to hypothesis to speculative theory. Some sections are technical; others draw on my own experiences, history and myth. I encourage you to read it as an initial introduction to the exploration of lifeforms that may be far stranger than we ever considered possible.

PART ONE

TOWARD A RECONCILIATION OF OBSERVATION AND SCIENCE

1 UFOs, UAPs and Other Lifeforms

When the New York Times ran their story on December the 16th 2017, titled "*On the trail of a secret Pentagon UFO program*", the landscape of the UFO phenomenon began to change. This was the beginning of a new era for ufology, where the public perception of Unidentified Flying Objects (UFOs) began to move from the marginalised fringe, into the spotlight of the mainstream media.

A follow up article in May 2019, titled "*Wow, what is that?*", and the subsequent release of three declassified US Navy videos on April 27th, 2020, known as the "*FLIR video*" (recorded 2004), the "*Gimbal video*" and the "*Go fast video*" (both recorded in 2015) brought the subject into sharp focus with the public.

In 2021, the term UFO was dropped in favour of UAP, which initially stood for Unidentified Aeriel Phenomena but was later redefined as Unidentified anomalous phenomena, so as to include a broader range of phenomena and distance the subject from its longstanding association with extraterrestrial craft and alien conspiracy theories.

Subsequent US congressional oversight hearings, along with whistleblower testimony, have sustained public interest, as an incremental push continues for disclosure of alleged extraterrestrial recovery programs and associated alien technologies. The declassified videos are intriguing, but fall well short of conclusive, "smoking gun" evidence of the extraterrestrial visitation.

Definitive proof, whether in the form of physical craft, recovered extraterrestrial biological material, or non-human technologies, that would conclusively end speculation and bring about genuine disclosure, remains elusive. As a result, the UFO and UAP subjects appear to be increasingly diffuse.

Many videos of anomalous phenomena are presented as solid, physical craft, even when the footage itself fails to support such a conclusion. Until tangible physical evidence is brought into the public domain for open, independent analysis and validation, witness testimonies and statements remain, by necessity uncorroborated, regardless of their origin.

Whistleblowers stop short of relaying any verifiable information, citing their fear of violating their Non-Disclosure Agreements (NDA) and sworn oaths. Regardless of intent, congressional oversight hearings have at times served as a highly visible media platform in which an already ambiguous subject has become more confused than clarified.

If a coordinated program of secrecy does exist, such conditions would be ideal for perpetuating confusion rather than resolution. After considering this landscape at length, I decided to investigate the phenomenon myself, relying not on conclusions drawn by others, but on direct observation and field research.

My aim was to clarify what, if anything, is moving unimpeded in our atmosphere, and over our land and oceans. If we take the alleged 1933 recovery of a UFO in Italy as the beginning of the coverup as is often cited, we have, by definition, a conspiracy that has already spanned over ninety years. In this context, my quest may sound ambitious.

However, sometimes progress can only be made by first questioning established assumptions. I believe the long-term engagement and secrecy surrounding the UFO/UAP subject points to four possible explanations, with respect to the officially released material and the most-cited military cases:

- The released videos depict physical craft and are connected to an alien presence on Earth, as has long been suggested.
- The released videos depict "*something else*", known or unknown by the US government and/or military, but not by humanity in general, which are being paraded as physical craft in an effort to continue a deception for reasons unknown.
- The released videos depict "*something else*", as yet unknown by humanity in general, including the US government and/or military.
- The fourth possibility I will cover in a later chapter, as I believe it deserves more than a couple of lines to explore fully.

My standpoint had always been in favour of the extraterrestrial hypothesis (ETH). I was convinced, as were many, by all the books, reports, documentaries and videos, that alien beings were here on earth and there existed physical evidence to prove it. Historically, public engagement with the subject has often attracted scepticism, and my own interest was no exception.

Discussion has often been informal or dismissive, despite the frustration caused by continued non-disclosure. I was confident that proof of the extraterrestrial hypothesis (ETH) would be forthcoming. I expected that decisive evidence supporting the extraterrestrial hypothesis (ETH) would eventually enter the public domain, and I followed the subject openly.

In this respect I have never hidden my views or shied away from the discussion. Due to the tangential direction my investigation has taken and the speculative hypotheses that follow, this line of investigation places the work outside prevailing interpretive frameworks.

My conviction, however, remains unchanged. I just want the truth to be out there for all who wish to know. My research and observations over the past few years have convinced me that many videos and reports are not of physical craft. At this early juncture, I would like to state that, in no way does the information in this book refute the extraterrestrial hypothesis (ETH).

It does, however, help to clarify many anomalous sightings. This book proposes that there is a non-human intelligence (NHI) present in our skies and oceans, with all the extraordinary abilities that are ascribed to many UFO and UAP reports. I hope to be able to show you in the pages that follow, why I believe this is the case.

In this first chapter, I will begin to lay out a hypothesis and offer descriptions of what, "something else", may be. This information will go against the grain of some readers; I say this because it also took me sometime to accept. I ask only that you withhold judgement, suspend disbelief, and remain attentive to alternative explanatory frameworks of the hidden biosphere and its inhabitants.

The following is an overview of observations of Unidentified Anomalous Phenomena (UAP) in Western Australia. This is by no means definitive or conclusive and will ultimately require further investigation and independent verification. We therefore begin with a simple but far-reaching question.

"What if Earth harbours a parallel biosphere, largely invisible not because it is alien, but because it is electromagnetic rather than biological?"

This question serves as the conceptual spine of what follows. It reframes UAPs not necessarily as alien visitors, but potentially as indigenous to the Earth. Lifeforms whose bodies are fashioned by plasma dynamics, whose behaviour reflects an electromagnetic ecology, and whose presence has been witnessed, interpreted, and mythologised across cultures, since the dawn of human history.

Recent advances in plasma physics have made the hypothesis of plasma-based intelligence more plausible. Laboratory studies on dusty plasmas have demonstrated that charged particles can self-organise into crystalline arrangements, propagate information-bearing waves, and maintain coherent structure under sustained electromagnetic input.

In the upper atmosphere, sprites, ELVES and transient luminous events reveal that Earth is already capable of producing large-scale, structured plasma environments.

If persistent electrical charge can be generated and maintained by mineral and crystalline substrates capable of sustaining electromagnetic resonance, then materials that store charge, concentrate fields, and seed persistent electrical gradients, may create stable environmental conditions. Conditions within which ordered, life-like plasma structures become physically admissible.

This book takes that possibility seriously and asks what kind of beings might arise under such conditions, and how they may have already shaped human perception, myth, and experience. Although I use the terms, "plasma entity", "plasma being", and "plasma-based lifeform" throughout this book, I wish to clarify that these labels and the theoretical framework they support are necessarily speculative in nature.

They are however, grounded in my own sustained observations and experiences, and represent what I currently believe to be the most coherent interpretation of the phenomenon. As previously stated, there is no independent verification for these claims, though it is my hope that others will expand upon this work through their own investigation and research.

The descriptions offered may therefore represent only a small cross-section of a much broader diversity of plasma-based life within a hidden electromagnetic biosphere.

Why this Hypothesis Matters Now

In recent years, the question of what inhabits our skies, oceans, and atmospheric boundaries has shifted from fringe speculation to a matter of legitimate scientific and governmental interest. Since 2017, when previously classified UAP encounters were publicly acknowledged by military and intelligence agencies, the narrative that unidentified anomalous phenomena (UAP) are simply misidentified aircraft or experimental technology has grown increasingly inadequate.

The release of sensor-confirmed incidents, objects without visible propulsion, demonstrating non-ballistic movement, has exposed a conceptual vacuum. Traditional extraterrestrial craft hypotheses struggle to explain the behaviour observed, while conventional meteorological explanations fail entirely.

Simultaneously, advances in consumer imaging are transforming the observational landscape. Modern low-light cameras, high-frame-rate mobile sensors, near-infrared night vision, and full-spectrum recording devices now exceed the sensitivity of scientific instruments from only a few decades ago.

This technological shift has democratised the sky; for the first time in history, ordinary observers can capture phenomena that previously went unnoticed or unrecorded. My own investigations in Western Australia form part of this broader transformation. High-quality footage revealing structured plasma-like forms is no longer a rarity, it is becoming increasingly common.

At the same time, a renaissance in plasma physics, atmospheric electricity, and complex systems theory has reopened scientific space for models of matter that behave in life-like, self-organising ways. Research into dusty plasma crystals, electromagnetic self-confinement, and field-stabilised structures suggests that coherent, dynamic, and even responsive plasmas can arise under natural conditions.

These developments create a scientific foundation that did not exist when earlier generations dismissed luminous entities as folklore or misinterpretation. Taken together, these shifts demand a new interpretive framework. The simplistic belief that unidentified objects are either extraterrestrial craft or misidentifications can no longer sustain the complexity of the evidence.

A hypothesis is required, capable of explaining behaviour, morphology, reactivity, and the clear interaction between environmental conditions and the luminous forms observed. It was within this context that I began regular observations using consumer low-light imaging equipment in Western Australia.

Initial Observations

Early observations of strange, luminous objects that were invisible to the naked eye, appeared when filmed in full spectrum. These objects flew, if that is the correct term as no wings were visible, with a rippling motion. These were not metallic craft high in the night sky, but were much smaller, moving not far above ground level. Tree height would provide an approximate altitude.

Strange, faint images were observed in the background "noise" of images and screenshots filmed in low light. Initially these were ignored as they were believed to be a quirk of the camera, artifacts or pareidolia, but over time similar forms were found on images from other cameras.

I decided that this was not an issue to be discounted if I was to progress with my quest. It had to be more than coincidence. The subtle lines and faint colour variations seemed to partly show coherent forms of strange "creatures".

There seemed to be so many, roiling and writhing like a shoal of fish at the surface of a pond while feeding, their forms also morphing during this process. To observe and isolate a single, well-defined, complete form was therefore difficult to achieve, as many were partly obscured. Many remained mostly unrecognisable, yet contained some subliminal biological analogue, rendering them inexplicably familiar.

Observed Morphology

A snout, nostrils, head, body and appendages. Some had no appendages and a serpent-like morphology and others with segmented, insect-like bodies, complete with legs and antennae.

Other forms were eerily reminiscent of dolphins and orca. Adjustment of images using Adobe Lightrooms various features, produced less obscured forms and proficiency increased with practice over time. These Cetacean-like entities are theorised to be the originators of luminous orb observations.

I appreciate that this may sound improbable. Had I not made these observations myself, I would be of the same opinion. This is why this subject needs further investigation, and more researchers to document additional evidence.

When filmed on a digital camera during daylight, the luminous area presents as a dark smudge. These smudges are hypothesised to be a separate type or species to those recorded in darkness, where they present as orange, elongated, and wingless.

Cameras and Framerate

Video footage is compiled from multiple consecutive images which give the illusion of movement when played back. A high framerate, which is the number of consecutive images recorded in one second, will display smoother, clearer movement but requires more light through the shutter and on to the sensor. This is preferable for fast moving objects.

A lower framerate allows more light through the shutter and on to the sensor allowing brighter, clearer images in lower light situations. This is less than ideal for fast moving objects, as the object will blur as it moves noticeably during each frame.

A luminous sphere for instance, if moving at high velocity in low light conditions, will display as an elongated form instead of its correct spherical shape. This can also be exacerbated in longer wavelengths such as infrared (IR) and near infrared (NIR), as the sensor requires larger receptors for these wavelengths. Less receptors means lower resolution.

The Sionyx Aurora pro has video framerates of sixty, thirty, twenty four, fifteen or seven and a half frames per second and a slo-mo option. Thirty or twenty four frames per second are used as standard in low light settings where streetlights are operational. Fifteen frames or less where less light is available would be used, however, I rarely film in complete darkness.

Sixty frames per second would be used when filming in daylight, although the Sionyx camera has rarely been used during daylight hours, where a recently purchased Sony α7Siii is preferred, or a Samsung S25 Ultra for convenience.

The elongation of the entity may be the subjects natural morphology, or may instead be caused by the speed of movement across the field of view of the camera. The elongation is not apparent during daylight recordings, which may be due to the higher frame rate of sixty frames per second UHD (ultra-high definition) which would greatly reduce this effect.

However, recordings using 8K resolution at thirty frames per second during daylight also reveal little, if any, elongation. If this observation is correct, it would suggest there are nocturnal and diurnal species. Inspection of full-spectrum recordings have shown these objects are initially visible at an estimated fifty metres from the camera.

There, they appear in frame as white or translucent. As they approach the camera they take on a green hue, and then a mottled orange when within approximately ten metres. The Sionyx Aurora Pro's night vision colour mapping is non-linear and proprietary, and is not a spectrometer. As such, colours are not accurate to our visible light spectrum.

For instance, green trees may appear as purple or pink. Therefore, for our close proximity orange entity, it is difficult to judge an accurate wavelength, although an orange to red tint usually correlates with NIR energy in the 850 nm – 950 nm range.

The entities leave a translucent trail on video when filmed at all distances, and when filmed close to the camera, leave an almost identical, faint, translucent version of the previous frames. This may be a ghosting effect from the previous frames but does not happen when insects are filmed, or it may be due to an electromagnetic interference effect but is strange in itself.

The entities movement is smooth and as if with purpose, often reappearing after a few minutes, if they are in fact the same entity, and covering a similar route, as if completing laps of an area or territory or searching for something. The following description is generally appropriate for the smaller, nocturnal type, and the larger type which tend to be at higher altitudes.

How distinct these types are, other than size, is currently unknown. Full-spectrum video observations of these smaller entities at close proximity, reveal the luminous part seems to comprise many small, cuboid shaped structures of varying colours on the exterior, giving it a bejewelled appearance.

The larger entities often present as stationary or slow moving, as if watching from above, often with a pulsing, multicoloured, luminous orb, which may glow, fade completely or even flash in just a few frames.

When adjustments are made within an image, further details may be exposed within the orb. Geometric shapes, such as square or rectangular cuboids. Analysis of thousands of video frames has led to the conclusion that the luminous orb is only a small, constituent part of a much larger entity.

The extended form comprises a finer, more subtle structure. Beginning with the snout-like appendage, situated immediately behind the luminous orb structure. The snout is often observed with two black spots where the snout and "head" intersect.

Observations of diurnal entity image captures, reveal the cuboid positioned internally. Observations of the internal cuboid were made through the open "mouth". With this type of entity, the snout is absent.

These smaller and larger nocturnal entities have a similar morphology. What appear to be eyes are situated laterally on the domed head, usually surround by a dark surface colouring. The reference to physical characteristics found on biological species is an assumption based solely on a biological analogue and may not be correct.

In terrestrial biology, predator species, such as leopards, have anterior arranged eyes, allowing binocular vision. This facilitates sharper focus and more accurate depth perception. Conversely, prey species such as antelope, have evolved laterally positioned eyes. This wider field of view favours an earlier warning of danger from predators and a greater chance of escape.

In both dolphins and orca, the eyes are laterally positioned, even though orca are apex predators. However, there is a small overlap in their field of view allowing limited binocular vision to home in on prey. Orca also use their eyes independently to scan more effectively for prey. How this relates to a plasma-based ecology is not known at this time.

The larger and smaller nocturnal entities have lateral, elongated, flipper type appendages which come from behind and below the eyes. Relative body scale between these two types, approximates that of dolphins to orca.

Their bodies taper less to the posterior than biological cetacean species, as there is no tail flipper. There is instead, something comparable to the rear flippers found on a seal or sea lion, ending in a set of what look like short, blunt, claws or toes.

Whales belong to the Artiodactyls, a suborder of even-toed ungulates and as such, have a pair of vestigial limbs at the rear of their body. The term vestigial means the limbs are non-functioning and are inside the body, a remnant of a time when the ancestors of whales were land based.

Whales are carnivorous unlike their land based cousins, which include hippopotamuses, cows, giraffes and camels, all of which are herbivorous. Pigs and peccaries are omnivorous, consuming both vegetation and meat and are the only known land based members of this order to do so.

Maybe it is not coincidental that there is a striking resemblance between the snout-like features of some plasma entities and the jade carvings of Pig Dragons (Zhu lóng), fashioned by the neolithic culture of the Hongshan (4700-2900BC), in north-eastern China and Mongolia.

On occasion, I have observed images containing a beard-like protuberance; long tendrils, maybe more like dreadlocks, which hang vertically in a mass from below the mouth area. I have seen videos which, like the black smudge discussed earlier, show only this feature during daylight, moving across the sky where it presents as a black or sometimes white mass of hanging tendrils.

These were described as: a jellyfish, a kite, a man in a jet pack and also a flying witch. These do resemble the various descriptions to a point although I don't believe we have yet discovered flying jellyfish, and the movement and speed are inconsistent with a kite.

With current technology, a jet pack is constrained by altitude and endurance limits of such equipment. A flying witch description probably stems from bias based on a religious motif or ideology.

In each case, such interpretations are inconsistent with the observed kinematics and endurance of the phenomenon. When viewed from the perspective of the plasma entity hypothesis (PEH), these sightings align closely with observations.

The refined outline of the entity's body structure may also be occasionally observed in Mie-scattered artificial light from streetlights. The scattering medium may be dust or minute water droplets from mist, and high humidity. When exposed, certain features of the entity can be made out.

Repetition of this process across many frames of video begins to build an overall image of these entities as just described. This description aligns with the entities observed with the naked eye. Although these sightings were rare and fleeting, they appeared translucent.

Biological Parallels

There is an obvious similarity between the description of these entities and terrestrial organisms. One has to wonder why this is. In no scenario can I imagine any evolutionary relationship between biological animals and these possibly plasma-based, interdimensional entities, yet the similarities are striking.

It may be coincidence, mimicry, or evidence of an underlying principle that shapes life in our universe or across dimensions. It is theorised that mimicking various aircraft, animals, birds, and potentially any number of animated objects is possibly part of a mating ritual, if of course, mating is applicable, as reproduction may be by other means.

Mimicry in nature is well documented; two examples of note are the mimic octopus and the superb lyrebird. The possibility of camouflage was also briefly entertained, though a plausible motivation was difficult to find for a mostly invisible entity.

Another possibility could be that constantly changing appearance through mimicry of other objects, is designed to confuse observers and resist categorisation. Is this the origin of myths and folklore of old? Maybe this behaviour is conceived to propagate new ones.

Another explanation for these similarities with biological life may simply be aerodynamic and hydrodynamic optimisation. Ocean-dwelling mammals do not travel more than a few metres in air, nor do birds swim far.

If these entities are trans-medium, meaning they travel in both air, water, and possibly also extraterrestrial environments, as is speculated, these biological parallels may just happen to be the perfect fit.

The Five "Observables"

The five original, and the more recently added sixth, observable conditions are required for a sighting to qualify as an exotic or alien craft and are often cited within the UFO community as proof of alien technology in our skies. However, plasma entities may also fulfil these conditions, which are briefly stated here:

1. Instant acceleration.
2. Hypersonic speed.
3. Low observability.
4. Trans-medium travel.
5. Anti-gravity.
6. Physical or invasive biological effects to close observers and the environment.

Whether these criteria are correctly interpreted as evidence of technology is debated; in this framework, they are treated as behavioural descriptors that may also apply to non-mechanical systems.

Acceleration, Velocity and Mass

Using the video frame rate and estimated distance, I have calculated speeds of thousands of kilometres per hour for some of these objects, sometimes from a standing start.

However, at this juncture, we can only speculate on how this is achieved. In chapter two we will take a closer look at the flight characteristics of these life forms and ways in which they may be realised with our current knowledge.

Instant acceleration causes extreme g-force on physical mass, which for a being constructed with a mass similar to ourselves would certainly be fatal, but the minimal mass of a plasma-based form is almost negligible.

During the infamous USS Nimitz Tic-tac incident of 2004, it was reported that a Tic-tac shaped object descended from 80,000 feet to sea level in 0.78 seconds. The object was reportedly confirmed on radar by a ~~United States~~United States Navy vessel.

For an object to travel 80,000 ft or 24.38 kilometres (15.15 miles), in 0.78 seconds is an average speed for that distance of Mach 91 (ninety-one times the speed of sound), or 30.9 km/second, 111,394 km/h (69,210 mph).

This cylindrical-shaped object was reported as a physical craft with no visible means of propulsion. I will not refute the statement that it was indeed a physical craft, as my knowledge of the incident is confined to media reports.

Suffice to say that many reports of physical UFOs may not

have been physical craft but were instead plasma-based entities. Regardless of size. The USS Nimitz incident may demonstrate that some reports of physical craft could in fact be misidentified plasma entities.

Hypothetical Mass Calculation

I would like to propose a hypothetical entity mass calculation and means of propulsion using Earth's magnetic forces as an option to the anti-gravity technology theories, often cited in many UFO and UAP reports. For the sake of our investigation in this book, let us look at a plasma-based entity with an orb/body complex of twelve metres, as in the USS Nimitz encounter.

If the twelve metre, cylindrical object was just the visible part of the entity, then let's also calculate mass for a ninety six metre long entity as could be scaled up using the proportions of an orca shaped body. What would the mass of such entities be?

To estimate the apparent mass, we will treat the mass as purely plasma. It might not have rest mass in the traditional sense, however, plasma consists of ionised particles, which are electrons plus nuclei, so there is still some particle mass.

Even under this extreme assumption, the mass remains trivial. With this in mind, we can very roughly estimate a minimum mass. This would equate to an entity, ninety-six metres in length having a volume of 19,302 m^3.

If we use the following plasma densities, we can calculate a mass range that these entities may fall within. Low plasma densities are typical of those found in auroras or diffuse atmospheric plasmas.

Low density plasma = 0.00001 kg/m^3

High density plasma = 0.001 kg/m^3

Using the following calculation

mass = ρ x V

where:

m = mass

ρ (rho) = density

V = volume

Twelve metre plasma entity

Volume: 37.7 m^3

Mass (Low): 0.000377 kg

Mass (High): 0.0377 kg

Ninety six metre plasma entity

Volume: 19,302 m^3

Mass (Low): 0.193 kg

Mass (High): 19.3 kg

Please note:

This is only a first order approximation. As such, any pressure, temperature, and exotic matter possibilities are ignored. If we treat the plasma volume as uniformly ionised gas, a simplification that ignores Debye screening and non-equilibrium effects, the resulting apparent mass is negligible.

We can see from these simple calculations that even a plasma entity ninety six metres long may have a mass as low as 193 grams, which is less than many mobile phones; if we base calculations on a twelve-metre-long entity, it could weigh the same as a grain of rice.

This low mass would allow an entity to have no significant weight or inertia, allowing high manoeuvrability at speed, which may be achieved electromagnetically. The behaviour is analogous to a system with negligible inertia, responding to minimal external forces.

A plasma entity, though vastly more complex, might share that lack of resistance, explaining its seemingly impossible manoeuvres. A low mass would also allow the entity to assume a translucent, ghost-like presence, potentially visible only under specific electromagnetic or optical conditions, and offer a high potential for field-based phenomena like electromagnetic interference, radar ghosting, and light emitted as a glow, as has often been reported.

This is of course hypothetical, and there are many types of plasma. These calculations do not account for variations in pressure or temperature. There is also the possibility the plasma may be composed of exotic matter, dark plasma, or other unknown states where mass could be functionally negligible or entirely different from baryonic mass.

The Orb Structure and Function

The larger plasma entities are often observed manifesting as an amorphous, glowing orbs, with the main body structure extremely faint or invisible. The orb or disc-like shape is typically located at the "mouth" area relative to its larger, more obfuscated body structure and is what is often cited as a craft.

When viewed in a video, the orb can often be seen as a constantly morphing cuboid, highly illuminated and seems to be an orb due to the bright glow emanating from the internal structure. This orb may serve as the anchor point, and the finer extended body may be a projection from this.

The extended body structure is theorised to be stabilised and modulated through the cuboid, internal, crystalline structure which may regulate plasma density for the entity. The luminescence of the orb is theorised to be controlled at will by the entity, reducing and increasing the emission, shape, and colour of light, or extinguishing it completely.

Larger entities appear to change the spherical orb shape into complex forms, mimicking aircraft and birds incredibly well, although not always perfectly. The mimicking of aircraft flashing lights may be inconsistent, or colours might vary, but to a casual observer this would usually not arouse suspicion.

This mimic behaviour was first noted in mid 2024 when aircraft were filmed one night in Perth's, Swan Valley. It was observed when reviewing the footage that the navigation light flash intervals from one particular aircraft were inconsistent, and that the object itself was indistinct and silent. Images were carefully scrutinised and faintly exposed the larger body structure of the entity.

When the New Jersey drone incursions began in mid-November that year, the realisation that they were not, in general, mechanical drones became apparent. They were similar in form to some of the entities observed in Western Australia.

Cuboid Exposure

In darkness, many strange, recurring flashes have been observed and filmed on both the full-spectrum Sionyx camera and a Sony α7Siii, a full-frame mirrorless camera. These flashes repeated after a few seconds without movement in the field of view.

The video frame containing the flash was isolated and cropped to increase the size of the flash, allowing closer inspection. After some adjustments, and occasionally without any adjustment at all, the subject can be clearly seen as a perfect geometric cuboid structure.

The possibility exists that these cuboid motifs might be artefacts of video discretisation, though it is intriguing that dusty plasmas can form ordered, cubic lattices under confinement. These cuboid structures can occasionally be seen to have faint lines through the cuboid form, dividing it into varying numbers of smaller cubes.

The quantity of smaller cubes making up the larger structure is theorised to indicate the possible development or longevity of the entity. Often, a glowing spherical or partly spherical structure can be seen overlapping the internal smaller cuboids, which fits within the main cuboid's six outer faces.

It is theorised that this spherical structure is the source of released electromagnetic energy, akin to neurons firing in the mammalian brain, as shown on an electroencephalogram. This was originally thought to be caused by low-resolution pixelisation of the image, but subsequent observations show smaller artefacts with curved edges in the image that are not pixelated.

Cuboid Complexity

These perfectly formed three-dimensional shapes often show that the cuboid composition consists of three or five rows of smaller cuboids. In the case of a cuboid with five rows repeated five times, ninety eight smaller cuboids surround an inner cuboid structure consisting of three rows of three cuboids, this gives twenty six smaller cuboids surrounding one cuboid at the centre.

The numbers 1, 27, 125 are cubed numbers, the sequence would continue 343, 729, 1331, and so on. The significance of this cubic scaling may be a representation of growth complexity in nested plasma shells, where each layer holds more charge, volume, density, or node complexity in the crystalline plasma matrix of the entity than the last, in a non-uniform way.

Plasma Crystal Arrangement

In complex dusty plasma physics, when micron-sized dust particles are suspended in an RF (Radio Frequency) plasma, they can become charged and arrange themselves into highly ordered plasma crystals. Under the right balance of electric, gravitational, and ion drag forces, the particles form a three-dimensional lattice.

Two of the most commonly observed arrangements are face-centred cubic (FCC) and body-centred cubic (BCC) cuboid or hexagonal structures, geometries which are also found in conventional metallic crystals. In an FCC structure, each particle is surrounded by its twelve nearest neighbours, resulting in a denser and more closely packed cuboid or hexagonal lattice.

In contrast, the BCC structure has only eight nearest neighbours, forming a more open framework. These structures are visible under laboratory conditions using laser sheet illumination and video microscopy, and they demonstrate that plasmas containing solid particulates can self-organise into stable, repeating geometric patterns in a crystal, much like atoms.

FRACTAL RECURSION

Fractal recursion is a blend of two ideas: Fractals are complex patterns that are self-similar, meaning if you zoom in, the smaller parts look like the whole, a repetition of a base pattern that repeats itself at different scales.

Recursion is a term used in mathematics and programming where a function calls itself, a repeating process where the solution to a problem depends on solutions to smaller instances of the same problem. Essentially, each smaller cube mirrors the whole, like a hologram or a fractal leaf pattern repeating itself.

These crystalline cuboid structures can be interpreted as forming into integer cube based layers, where each layer contains a specific number of charged particles. In theory, this dense plasma area of the entity could be holographic in nature, and if fractal recursion was a utilised mechanism, this could allow a blueprint of the entity as a whole to be contained in each cuboid section, just as in a hologram, effectively holographic DNA.

Information about the entity's global structure could, in principle, be distributed non-locally through self-similar field oscillations, analogous to holographic storage If the cuboid structure is the controlling mechanism of the entity, then fractal recursion would allow non-localised consciousness, whereby if part of the entity became damaged, it would still be able to function and would allow for self-repair or healing, and self-replication.

Video footage captured by an MQ-9 Reaper drone on the 30th of October 2024, often referred to as the "Yemen orb" footage, was later presented publicly in 2025. The recording depicts a high-speed object struck by a ~~UNITED S~~United States~~TATES~~ military AGM-114 "Hellfire" missile. Although the impact appeared to fragment part of the object, it continued on its trajectory without evident catastrophic loss of coherence.

One speculative interpretation is that fragmentation did not constitute destruction but structural division. If the object represented a plasma-based system organised through non-localised or distributed architecture, such division could conceivably permit continued function or even replication.

A biological analogue can be found in organisms such as hydra, which can regenerate a complete organism from a fragment because structural and informational organisation is distributed throughout the body.

Dusty Plasma Crystalline Structures

In standard plasma physics, solid crystalline structures and plasma do not usually coexist, because high-energy plasma tends to disrupt or vaporise conventional solids.

However, it seems that in advanced plasma research and speculative physics, some concepts bridge that gap. So far, it has not been determined if there are more layers within each of the smaller cubes, but as this is a theoretical, sentient, plasma entity model, let us test its limits.

The crystalline cuboid is theorised to be a dusty plasma crystal, where micron or nano sized grains of matter are charged and suspended in plasma. These particles self-organise into geometric lattices due to Coulomb forces, also known as electrostatic forces. These forces are the attractive or repulsive forces between electrically charged particles.

A simple way to picture this is to think of tiny, charged dust grains behaving like magnets in a box. Some snap together, others push apart, but collectively they fall into neat, repeating patterns. Plasma performs this same dance on an invisible stage, arranging itself into lattices that seem almost crystalline.

These charged particles may form the cuboid lattice which could be constantly restabilised by electromagnetic forces from the plasma sphere or shell within the cuboid, to maintain the cuboid shape even in turbulent environments.

If this is the case, could the spherical plasma shell within the cuboid structure, function to effectively pressurise the cuboid structure from within?

Electrostatic Attraction and Repulsion

The presence of the inner sphere offers two plausible modes of stabilisation for the cuboid lattice. The first rests on electrostatic attraction. If the sphere carries a charge opposite to that of the surrounding cuboid structure, then Coulomb forces will draw the two into mutual balance.

Each face of the cuboid would feel an inward pull towards the charged core, helping to preserve its form against dispersal. In this way the sphere acts almost like a binding agent, anchoring the larger lattice around itself.

A second possibility lies in symmetry of repulsion. If the inner sphere and the cuboid share the same charge, then direct attraction is impossible. Instead, each face of the cuboid experiences an equal repulsive force radiating outward from the core.

Rather than collapsing or flying apart, the cuboid may remain suspended in equilibrium, its symmetry maintained by the balanced outward pressure. This situation has parallels in plasma confinement, where charged shells can hold form through a fine equilibrium of opposing forces.

In both cases the sphere functions as a stabilising element, whether by drawing the cuboid inward through attraction or by maintaining structural balance through repulsion. The precise mechanism may depend on charge distribution within the entity, but either pathway suggests that the inner core is indispensable for the coherence of the larger structure, and both may occur dynamically, depending on charge shifts.

This attraction and repulsion is not unlike the way balloons stick to a wall after being rubbed on hair, or how two charged rods repel each other. At a microscopic scale, these pushes and pulls could become the architecture of entire plasma beings.

We have to wonder, why a cuboid? In nature, plasma tends towards spherical or toroidal symmetry as these shapes reduce energy loss and increase stability, but the cuboid shape may serve a unique function. This unusual set-up containing a sphere makes it an interesting proposition for a theoretical model.

If this structure is indeed a three-dimensional, multi-layered, fractal lattice, the geometry could allow the cuboid structures to resonate at specific frequencies, similar to a cavity resonator.

Cavity Resonators

A cavity resonator is a structure typically made of a conductive material that confines electromagnetic waves, usually at dimensions comparable to the wavelength of electromagnetic energy in free space and allows them to oscillate at specific resonant frequencies. An analogy would be the body of a violin.

The wooden chamber shapes and amplifies the vibrations of the strings into a resonant tone. In the same way, cuboid cavities within plasma entities could be shaping electromagnetic waves, turning raw energy into coherent resonance.

Cavity resonators may be any geometric shape but are typically, and conveniently, rectangular cuboids. In the case of our plasma entity, this resonance could stabilise plasma oscillations.

Layered Three Dimensional Structures

The edges of the cuboid may precisely focus electromagnetic fields in specific directions, aiding in gravitational and magnetic manipulation for motion and manoeuvrability. The one hundred and twenty-five smaller cubes, layered to make up the larger three-dimensional cuboid structure, may each be made up of one hundred and twenty five smaller cubes, and so on.

As observations suggest an overall size and scale of a typical plasma entity to be that of an orca, this would give an orca brain volume of 0.006859 cubic metres or 6859 cubic centimetres, so for our theoretical entity brain, this is our basis for calculation.

As previously stated, I believe there are much larger entities. If we take my earlier suggestion of a cavity resonator and match electromagnetic wavelengths to the size of each cube, scaled down layer by layer, we get the following figures calculated to ten layers, as shown in Table 1.

Layer	Cavity Size	Type of EMF	Number of Cubes
1	19.00 cm	UHF	1 cube
2	~3.79 cm	Microwave	125 cubes
3	~7.58 mm	Infrared	15,625 cubes
4	~1.52 mm	Infrared	1,953,125 cubes
5	~0.30 mm	Infrared	244,140,625 cubes
6	~6000 μm	Infrared	3.05 x 10^10 cubes
7	~12.00 μm	Infrared	3.81 x 10^12 cubes
8	~2.40 μm	Visible light	4.77 x 10^14 cubes
9	~485.0 nm	Ultraviolet	5.96 x 10^16 cubes
10	~97.00 nm	X-rays	7.45 x 10^18 cubes

Table 1

These figures include reduction in resonant cavity size down to the UV wavelength range. Resonant cavities smaller than approximately 150 nm would struggle to sustain higher frequencies, which face increasing plasma decoherence.

Therefore, the UV wavelength is a plausible boundary for coherent electromagnetic resonance in a plasma entity, with longer wavelengths being more natural for energy efficiency.

Comparing the Plasma "Brain"

When the total number of resonant domains in the ultraviolet range is compared with the estimated, eighty-six billion neurons in the human brain, the resulting disparity highlights a form of complexity that defies biological intuition. Within a volume of approximately 0.19 m^3, the theoretical cuboid plasma brain contains an extraordinary number of electromagnetic resonators across its nested layers.

Although these domains cannot be equated directly with neurons, since resonators are modes rather than discrete cells, the sheer quantity involved indicates a computational density far beyond that of neurobiological structures. This model does not imply a one-to-one functional correspondence between resonators and neurons.

A resonant cavity is not a biological cell but an oscillatory region that can sustain multiple electromagnetic modes simultaneously. Long-wavelength modes span the entire volume, while progressively higher frequency modes occupy increasingly fine subdivisions.

Each region of plasma therefore participates in several modes at once, generating a hierarchy of interacting frequencies. The total number of potential oscillatory units, when taken across all scales, exceeds the number of neurons by many orders of magnitude, even though the functional analogy is conceptual rather than literal.

Such density suggests a potential for information storage, pattern formation and parallel processing that surpasses what is achievable by biological brains. Neurons operate through electrochemical pulses that are constrained by synaptic bottlenecks and metabolic limitations. By contrast, resonant plasma domains may operate at electromagnetic frequencies that are millions to billions of times faster than synaptic signalling.

This permits the possibility of ultra-fast cognition, rapid reconfiguration of internal states and a form of perception that is directly coupled to environmental electromagnetic phenomena. At the smallest scales, cavity dimensions approach the nanometre range, where classical and quantum behaviours overlap.

Quantum coherence, phase stability, and field entanglement, understood here as sustained electromagnetic coupling and phase-correlated interaction across nested resonant domains, rather than strict nonlocal quantum entanglement, may therefore contribute to the organisation of these structures.

The fine granularity of resonant domains would allow such a being to sense, interpret, and manipulate external fields with a level of precision wholly beyond human capability. In effect, the plasma brain may function as a spectral architecture, with cognition emerging from the interplay of harmonics, interference patterns, and nested resonances.

This points to a form of intelligence that is radically alien to human cognition. It is unlikely to be linear, sequential or dependent upon symbolic language. Instead, it may be distributed throughout the entity, holographic in structure and multidimensional in operation.

Its coherence may emerge from the synchronisation of frequencies across scales, from ultraviolet to UHF, producing internal order through field interactions rather than through neural connectivity.

The cuboid crystalline structure proposed in this model may stabilise its form through nested resonators and internal field geometry, creating a level of organisation beyond any known state of plasma in terrestrial science.

A Case for Non-Human Intelligence

The implications of this nested resonant architecture are profound. If such a structure functions as a computational system, it would represent a conscious intelligence that is neither biological nor mechanical. Its awareness would be distributed across trillions of oscillatory domains rather than confined to discrete cells.

Its cognition would arise from the coherence of electromagnetic fields, not from the sequential firing of chemical synapses. Such an entity may process information in parallel across multiple scales of frequency and resonance, producing a form of thought that is capable of instantaneous reorganisation and near-limitless integration of sensory data.

This form of intelligence would be fundamentally different from the human mind. It would not rely on discrete neuronal signalling, but on continuous field interactions. It would not think in sentences or images, but in harmonic patterns and spectral shifts. It would not perceive its environment through slow, receptor-based pathways, but through a direct electromagnetic coupling.

What appear to us as flashes of light, rapid changes of colour or shape, or the mimicry of familiar forms may represent expressions of communication, states of attention or emotion, or complex functions that lie outside our conceptual framework.

If the cuboid crystalline plasma model is correct, such beings would embody a mode of organisation that is non-biological yet highly ordered. Their intelligence would be fluid, distributed and non-local, grounded in the physics of resonance rather than in the chemistry of life.

They would constitute a distinct form of mind, incomparable to human cognition in structure but perhaps comparable in capability, or even surpassing it by vast margins. This raises the possibility that the luminous entities observed throughout history belong to a parallel evolutionary lineage, one rooted not in carbon, but in the dynamic architecture of electromagnetic fields.

Conclusion

This chapter has theorised how plasma-based lifeforms, due to their almost negligible mass and subtle body structure, may be able to achieve incredible acceleration and reach extreme velocities, while remaining mostly hidden.

Their intelligence may be greater than our own by orders of magnitude, and that their thought processes may not be encumbered by the slow processes that limit biological life.

In the next chapter we shall discuss in more detail, possible means for the incredible movement abilities of these entities and how they align with established criteria for distinguishing unidentified anomalous phenomena (UAP) from recognisable, manufactured objects and known biological systems.

2 Movement of a Species

If many UAP sightings are indeed plasma-based lifeforms, as is theorised, then the exotic explanations involving anti-gravity or warp-drive technology become unnecessary. Instead, we may consider alternative mechanisms rooted in electromagnetic field interactions, plasma dynamics, and natural atmospheric processes.

In simple terms, a technology, by definition, is the deliberate application of scientific knowledge to design tools, systems, or processes that make tasks achievable, easier, or more efficient. By implication, technology presupposes manufacture or deliberate modification, a distinction that becomes important when considering naturally occurring or biologically emergent plasma systems.

There may be a case for plasma entities being manufactured lifeforms, however, for our current model we will assume that this is not the case, although I do not rule this out. Let us now examine the criteria commonly used to characterise anomalous aerial phenomena within contemporary UAP analysis, most notably those associated with the Advanced Aerospace Threat

Identification Program (AATIP) and related investigative efforts.

These criteria are often referred to collectively as the "Five Observables". The original five observables are listed below. In some analyses, a sixth category, relating to physiological or biological effects, is also included. While not always formally grouped with the original five, it is relevant to later discussion in this book and is therefore included here for completeness.

Of these observables, four are primarily concerned with motion and kinematics. This chapter will focus chiefly on reconciling those four movement-related characteristics within the framework of plasma-based lifeforms. These observables function not as proof thresholds, but as analytic markers used to distinguish genuinely anomalous behaviour from conventional aerospace or atmospheric phenomena.

These follow the physical, observable constraints used to characterise anomalous phenomena:

1. Instant acceleration refers to the observation of extremely high, near-discontinuous changes in velocity or direction by an object, occurring over timescales too short to be compatible with known propulsion technologies or inertial tolerance limits. Such manoeuvres appear to exceed the structural, thermal, and biomechanical thresholds of all conventional vehicles and living organisms, and are not accompanied by observable exhaust, control surfaces, or reaction forces.

2. Hypersonic speed refers to velocities exceeding Mach 5, a regime in which airflow around an object becomes dominated by strong shockwaves, extreme aerodynamic heating, and highly compressed boundary layers. In the

context of anomalous aerial phenomena, the term additionally denotes the observation of such speeds without the accompanying sonic booms, thermal signatures, ionisation trails, or structural shock interactions required by conventional aerodynamics.

3. Trans-medium travel refers to the capacity of an object or phenomenon to traverse multiple distinct physical media, including atmospheric gases, liquids, and the vacuum of space, while maintaining stable operation, structural coherence, and consistent performance characteristics. Such behaviour occurs without observable mechanical adaptation, propulsion changeover, or degradation of motion despite dramatic shifts in density, pressure, drag, refractive index, or electromagnetic conditions.

4. Anti-gravity is a colloquial term used to describe the apparent ability of an object or phenomenon to counteract or overcome gravitational forces without the use of aerodynamic lift, mechanical thrust, buoyancy, or reaction-mass propulsion. In reported anomalous cases, this includes stable levitation, rapid vertical ascent, and controlled motion that occurs without observable exhaust, pressure waves, or interaction with the surrounding medium.

5. Low observability refers to the reduced detectability of an object across multiple sensing domains, including visual, infrared, acoustic, and radar spectra. An object exhibiting low observability may appear intermittently, display low contrast against its background, produce minimal thermal or acoustic signatures, and evade conventional radar returns. The phenomenon implies either deliberate signature suppression or an intrinsic physical state that

weakens interactions with electromagnetic or acoustic detection methods.

6. Physical or invasive biological effects refer to measurable physiological, neurological, or environmental impacts associated with close-range encounters with UAP. These may include heat exposure, skin irritation, temporary paralysis, nausea, altered cognitive states, electromagnetic interference, damage to vegetation or infrastructure, and other anomalies not attributable to known environmental causes. Such effects imply that the object or phenomenon is emitting energy or fields capable of interacting with biological systems and physical materials.

Environmental Correlations

Plasma lifeforms may find a natural habitat within charged atmospheric structures. Thunderstorms, ionised cloud layers, and geoelectrical boundaries provide ideal energetic substrates for movement, feeding, or communication. Lightning, in particular, offers a vast reservoir of electrical potential, and may serve as both a navigational aid and a source of sustenance for these entities.

The frequent correlation between observed unidentified anomalous phenomena (UAP) and storm activity suggests a dependency upon, or at least an affinity for, highly charged atmospheric environments. Another long-term series of strange occurrences which has resisted a definitive explanation is the *Sky Trumpet phenomena.*

We will address this in a later chapter in detail, but I will briefly mention it here. Described in many reports as low-frequency mechanical hums or metallic groaning sounds, one possibility for this age-old, cross cultural phenomena may be a direct consequence of Electroaerodynamic (EAD) propulsion cycles or internal modulations of resonant plasma structures.

Emissions in the range of 20 Hz - 150 Hz could result from standing waves established within large-scale plasma entities, possibly interacting with urban infrastructure, geological strata, or atmospheric ducts.

Magneto-Optic and Visibility Effects

The fins may also function in visibility modulation. By altering local magnetic field strength, a plasma entity may be able to shift the refractive index of the surrounding medium. Magneto-optic effects, such as the Faraday rotation, could create transient cloaking, chromatic lensing, or shimmer-like distortions.

Combined with electroaerodynamic (—EAD)-induced boundary layers, this would explain the common reports of entities that vanish, flicker, or appear semi-transparent. These optical effects may be consciously controlled, or side effects of motion, suggesting a tight coupling between propulsion, visibility, and internal field regulation.

Propulsion Possibilities

This chapter combines two theoretical models concerning mechanisms by which plasma-based lifeforms may interact with the environment and navigate through water, the atmosphere and beyond. The first examines the principles of electroaerodynamic (EAD) propulsion, a form of movement generated through ionised airflow manipulation.

The second explores the dynamics of lateral angular fins, hypothesised to interact with Earth's magnetic field to facilitate movement, sensory perception, and field-based communication.

Together, these models offer a cohesive framework for understanding the motion, manoeuvrability, and perceptual systems of plasma-based entities as observed in anomalous aerial phenomena.

Electroaerodynamic Propulsion

Electroaerodynamic propulsion is a method of generating thrust without moving parts. It is based on the principle that when a high voltage is applied between an emitter and collector electrode, the surrounding air becomes ionised.

The resulting ions accelerate towards the oppositely charged electrode, dragging neutral air molecules along with them, thereby creating ionic wind and net thrust. This mechanism has been experimentally demonstrated in devices such as lifters, which are capable of levitating small structures using nothing but high-voltage power supplies.

While efficient at small scales, the thrust-to-power ratio and dependency on atmospheric conditions have traditionally limited electroaerodynamic (EAD)'s applicability in conventional aviation.

However, in plasma-based lifeforms composed of ionised matter, electroaerodynamic (EAD) principles may operate with far greater efficiency and dynamism. If entities are capable of internally generating or modulating electric fields through the movement of charged plasma within cavity resonator structures, these internal field geometries would allow for precise control of local ion flows and airflow propulsion across their surface.

Magnetic Interactions

Beyond EAD propulsion alone, plasma entities could interact with the Earth's magnetic field via Lorentz force coupling. The movement of charged particles within a magnetic field produces orthogonal forces, and this principle could be exploited to produce lateral or vertical motion with minimal apparent inertia.

The degree of mobility observed in many aerial phenomena, such as instantaneous acceleration, silent hovering, and right-angle turns, could be facilitated by such magneto-plasma dynamics. If these lifeforms possess internal mechanisms for generating or modulating magnetic fields, they may couple with the ambient geomagnetic environment to achieve inertial damping or localised spacetime deformation.

In effect, their presence in the atmosphere may constitute not merely a visible or energetic disturbance, but a local reorganisation of electromagnetic potential fields. Such a capability might explain their high manoeuvrability and intermittent radar returns, as well as their apparent invisibility at certain frequencies.

The Lorentz force

The Lorentz force is the fundamental physical interaction that governs how charged particles and current-carrying plasmas behave in electric and magnetic fields. When a particle possesses an electric charge, it experiences a force whenever it moves through a magnetic field or encounters an electric field. This force determines its acceleration, direction of motion, and the overall dynamics of plasma structures.

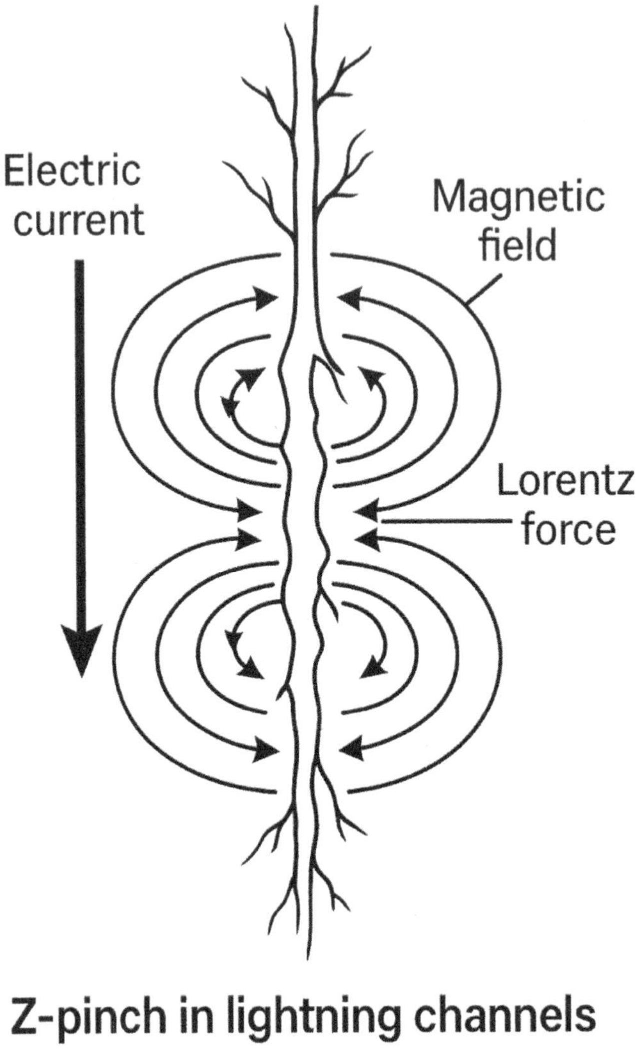

Fig. 1. Fork lightning Z-pinch effect

In its simplest form, the Lorentz force acts in two components: an electric component, which pushes the charge directly along the direction of the electric field, and a magnetic component, which acts at right angles to both the motion of the particle and the magnetic field.

The combined effect can cause charged matter to curve, spiral, oscillate, or become trapped within field configurations, allowing plasmas to form stable shapes, filaments, or rotating shells. In natural environments such as the ionosphere, auroral arcs, lightning channels, and magnetospheric currents, Lorentz forces play a central role in organising plasma behaviour.

For a plasma-based entity, the Lorentz force would provide an efficient means of generating lift, steering, acceleration, and trans-medium manoeuvring by continuously shaping the motion of charged particles within its structure. When we observe fork lightning during a thunderstorm, this is the Lorentz force in action.

The Lorentz Force constricts the current into a narrow, filament-like structure through a process known as magnetic pinching or the "Z-pinch effect". When a large electric current flows through plasma, it generates a magnetic field that encircles the current path. This magnetic field produces an inward-directed Lorentz force on the charged particles within the plasma column.

As the current increases during a lightning discharge, often tens of thousands of amperes, the surrounding magnetic field strengthens proportionally, tightening its grip on the ionised channel. The result is a self-compressing plasma filament only a few centimetres wide, despite carrying immense electrical power.

This magnetic constriction explains why lightning channels remain remarkably narrow and coherent over long, tortuous paths. In the context of plasma-based entities, a similar

Lorentz-driven confinement mechanism could maintain body coherence, create internal structure, or support rapid changes in shape or propulsion.

Geomagnetic coupling

Geomagnetic coupling describes the process by which a plasma structure interacts dynamically with the Earth's magnetic field, allowing energy, momentum, and information to flow between the entity and the geomagnetic environment.

Because plasma consists of charged particles, it naturally responds to magnetic fields: as it moves, each charged particle experiences a Lorentz force that can alter its trajectory, confine it, or accelerate it along preferred directions.

In Earth's magnetosphere and upper atmosphere, this coupling governs the behaviour of auroral arcs, lightning leaders, sprites, and other ionised structures that align with or bend according to geomagnetic field lines.

For a plasma-based entity, geomagnetic coupling would allow it to anchor itself to ambient field lines, draw energy from large-scale ionospheric currents, or convert magnetic tension into controlled motion.

By adjusting its internal charge distribution or current pathways, the entity could strengthen or weaken its interaction with the geomagnetic field, enabling lift, stability, steering, or even rapid acceleration without mechanical propulsion.

This interaction provides a plausible physical framework for silent levitation, sharp turns, and trans-medium movement, as the entity would be manipulating magnetic forces rather than pushing against air or water.

Angular Fins as Magnetic Field Modulators

The second component of this dual-propulsion model centres on the use of lateral angular fins, translucent extensions that may function as magnetic field manipulators. These fins could serve as dynamic field-control structures, similar to vectored thrust nozzles in jet propulsion but operating through geomagnetic coupling.

Plasma, being electrically conductive, responds to magnetic fields through the Lorentz force. The fins' shape, curvature, and charge distribution may be modulated in response to Earth's magnetic field gradients.

This would enable rapid shifts in orientation, angular momentum control, and lift vectoring without mechanical force application. Fin asymmetry could also generate torque, explaining observed spiral flight patterns and rolling behaviour.

Power Requirements

In chapter one, we established that a plasma entity of twelve metres length could possess an apparent mass as low as 0.000377 kg (low-density auroral plasma) or as high as 0.0377 kg (dense atmospheric plasma).

A ninety-six-metre entity would span a mass range from approximately 0.193 kg to 19.3 kg under the same assumptions. Even under the high-density model, the mass of such an entity is extraordinarily small relative to its geometric scale. T

his section develops the implications of that low mass for acceleration, manoeuvrability, and the hypothetical power requirements of an electroaerodynamic or electromagnetic propulsion mechanism.

Kinetic Energy Requirements for Hypersonic Travel

If we consider the entity as a body of ionised gas, isolated from aerodynamic interaction for the moment, then the mechanical energy required to accelerate it to a hypersonic velocity v is given by the classical expression:

$$E\_kin = (1/2) \times m \times v^2$$

where:
E_kin = kinetic energy (joules)
m = mass (kilograms)
v = velocity (metres per second)

Choosing a representative hypersonic velocity of $v = 3000\ m \cdot s^{-1}$ (approximately Mach 9 – 10 at stratospheric altitudes), the resulting kinetic energies for the four mass cases are:

Twelve-metre entity, low density **(0.000377 Kg)**:

E_kin = (1/2) × 0.000377 × (3000) ^2

E_kin = 0.5 × 0.000377 × 9,000,000

E_kin ≈ 1,696.5 J

E_kin ≈ 1.7 × 10^3 J

Twelve-metre entity, high density (0.0377 Kg):

$$E_{kin} = (1/2) \times 0.0377 \times (3000)^2$$

$$E_{kin} = 0.5 \times 0.0377 \times 9{,}000{,}000$$

$$E_{kin} \approx 169{,}650 \text{ J}$$

$$E_{kin} \approx 1.7 \times 10^5 \text{ J}$$

Ninety-six-metre entity, low density (0.193 Kg):

$$E_{kin} = (1/2) \times 0.193 \times (3000)^2$$

$$E_{kin} = 0.5 \times 0.193 \times 9{,}000{,}000$$

$$E_{kin} \approx 868{,}500 \text{ J}$$

$$E_{kin} \approx 8.7 \times 10^5 \text{ J}$$

Ninety-six-metre entity, high density (19.3 Kg):

$$E_{kin} = (1/2) \times 19.3 \times (3000)^2$$

$$E_{kin} = 0.5 \times 19.3 \times 9{,}000{,}000$$

$$E_{kin} \approx 86{,}850{,}000 \text{ J}$$

$$E_{kin} \approx 8.7 \times 10^7 \text{ J}$$

Even the largest and densest hypothetical entity requires only around 10^8 joules to reach extreme hypersonic speed, comparable to the energy in a few litres of aviation fuel or a small lightning discharge. At the low-density end of the spectrum, the kinetic energy becomes almost trivial.

Timescales and Power Output

Power is the rate at which energy is delivered:

$$P_{mech} = E_{kin} / t$$

where:

P_{mech} = mechanical power (Watts, W)

E_{kin} = kinetic energy (Joules, J)

t = time over which acceleration occurs (seconds, s)

If the acceleration occurs over different characteristic timescales, the required mechanical power varies dramatically.

For the ninety-six-metre, high-density case (19.3 kg):

Acceleration to 3000 $m \cdot s^{-1}$ in 10 s:

$$P_{mech} = (8.7 \times 10^7 \text{ J}) / 10 \text{ s}$$

$$P_{mech} = 8.7 \times 10^6 \text{ W}$$

$$P_{mech} \approx 8.7 \text{ Mw}$$

Acceleration to 3000 $m \cdot s^{-1}$ in 1 s:

$$P_{mech} = (8.7 \times 10^7 \text{ J}) / 1 \text{ s}$$

$$P_{mech} = 8.7 \times 10^7 \text{ W}$$

$$P_{mech} \approx 87 \text{ MW}$$

Acceleration to 3000 $m \cdot s^{-1}$ in 0.1 s:

$$P_{mech} = (8.7 \times 10^7 \text{ J}) / 0.1 \text{ s}$$

$$P_{mech} = 8.7 \times 10^8 \text{ W}$$

$$P_{mech} \approx 0.87 \text{ GW}$$

For the low-density case of the same size, these figures are reduced by a factor of 100. For the smaller twelve-metre entity, the power requirements are reduced by several orders of magnitude again; in the ultra-low-density case they become almost negligible in engineering terms.

From Mechanical Power to Field-Based Propulsion

If electroaerodynamic or electromagnetic propulsion mechanisms are invoked, the mechanical power must be supplied by a deeper energy reservoir. For a field-driven mechanism, the relevant relationship is:

$$P_{elec} = P_{mech} / \eta$$

where:

P_{elec} = electrical or field power input (Watts, W)

P_{mech} = mechanical power output (Watts, W)

η (eta) = efficiency of converting electrical or field energy into bulk kinetic motion

Human-built electroaerodynamic systems exhibit very low efficiencies, typically η in the range 0.01 to 0.1

Using the optimistic upper bound of $\eta = 0.1$:

A ninety-six-metre, high-density entity accelerating to $3000\ \text{m}\cdot\text{s}^{-1}$ in one second would require approximately:

P_elec = 8.7 x 10^7 / 0.1

P_elec = 8.7 x 10^8

P_elec = 870 MW

The same entity accelerating in 0.1 s would require:

P_elec = 8.7 x 10^8 / 0.1

P_elec = 8.7 x 10^9

P_elec = 8.7 GW

Interpretation

While these power levels far exceed the capabilities of conventional aircraft or drones, they fall well within the energetic range of natural plasma systems.

Thunderstorms, lightning discharges, auroral currents, and magnetospheric flux tubes routinely involve instantaneous power levels ranging from megawatts to gigawatts, with brief transient peaks reaching higher levels in lightning discharges. For a plasma-based organism capable of coupling directly to environmental electromagnetic fields, such energy demands would not be extraordinary.

At the lower end of the mass spectrum, such as a twelve-metre, low-density plasma entity, the required mechanical and electrical power would drop by several orders of magnitude, potentially into the kilowatt or low-megawatt range.

This reinforces the plausibility that extreme manoeuvres may arise naturally from plasma-field interactions rather than from engineered propulsion systems.

THE AERODYNAMIC DRAG CAVEAT

The calculations above treat the plasma entity as if it were accelerating in a vacuum. In reality, hypersonic flight is dominated by shock-layer formation and extreme aerodynamic drag. The power required to overcome drag at a given speed can be expressed as:

$$P\_drag = 0.5 \times \rho \times v^3 \times C\_D \times A$$

where:

P_drag = power required to overcome drag (Watts, W)

ρ (rho) = air density (kg/m^3)

v = velocity ($m \cdot s^{-1}$)

C_D = drag coefficient (dimensionless)

A = reference (cross-sectional) area (m^2)

This relationship differs fundamentally from the kinetic energy equation because mass does not appear. Even an object with negligible mass will experience enormous drag power requirements if it interacts with air like a solid body at hypersonic speed.

Two distinct interaction regimes therefore present themselves:

1. Solid-Body Aerodynamic Coupling

   In this regime, the entity interacts with air like a conventional physical object. Shock layers form, drag dominates, and the power required to maintain hypersonic velocity rises into the hundreds of megawatts or even gigawatts, even for an effectively massless plasma shell.

   The entity must expend immense energy to force a channel through dense air, resulting in heating, compression, and shockwave formation. This behaviour is inconsistent with many observed UAP characteristics, such as the absence of sonic booms, thermal signatures, and visible shock fronts.

2. Field-Decoupled or Plasma-Sheath Motion

   In this scenario, the entity does not couple directly to the surrounding air. Instead, it may generate a low-density plasma cavity, a magnetically shaped boundary layer, or a dynamically adjustable electromagnetic "slip" surface that suppresses shock formation.

   Under such conditions, effective aerodynamic drag could be reduced by orders of magnitude, allowing hypersonic manoeuvres to occur with the more moderate mechanical and electrical power levels calculated earlier.

This distinction, solid-body drag versus field-decoupled drag, is central to understanding why low-mass plasma entities could perform manoeuvres that appear impossible under classical aerodynamics.

Electromagnetic Propulsion via Earth's Magnetic Field

To evaluate whether the Earth's magnetic field could accelerate or steer such an entity, we consider two classic Lorentz-force mechanisms:

Motion of a charged body through a magnetic field

$$F = q \times v \times B$$

A current filament or loop interacting with the geomagnetic field

$$F = I \times L \times B$$

where:

F = force in newtons (N),

q = net charge in coulombs (C),

v = velocity in metres per second ($m \cdot s^{-1}$),

B = magnetic field strength in tesla (T),

I = current in amperes (A),

L = the effective length of the current path in metres (m).

Near Earth's surface, the geomagnetic field strength is approximately:

$B \approx 30\text{–}60$ microtesla

$B \approx (3\text{–}6) \times 10^{-5}$ T

At altitudes between approximately 12 and 20 kilometres, this value changes only modestly and remains of the same order of magnitude. This constancy allows geomagnetic coupling mechanisms to operate across a wide range of atmospheric heights.

Propulsion by Net Charge

The Lorentz force acting on a charged body moving through a magnetic field is given by:

$F = q \times v \times B$

To accelerate a ninety-six-metre, high-density plasma entity (19.3 kg) with a moderate acceleration $a = 10\ \text{m·s}^{-12}$, the required force is:

$F = m \times a$

$F = 19.3\ \text{kg} \times 10\ \text{m·s}^{-12}$

$F \approx 193$ N

Assume a representative manoeuvring velocity of:

$v = 1000\ \text{m·s}^{-1}$

and a representative geomagnetic field strength of

$B = 5 \times 10^{-5}$ T

Solving the Lorentz force equation for the required net charge q:

$$q = F / (v \times B)$$

$$q = 193 / (1000 \times 5 \times 10^{-5})$$

$$q = 193 / 0.05$$

$$q \approx 3.86 \times 10^3 \text{ C (Coulombs)}$$

This corresponds to a net charge of approximately 4,000 coulombs.

Such a charge would be utterly impractical for a rigid metallic craft, but it is not extreme for plasma systems. Natural plasmas routinely sustain charge separations ranging from 10^6 to 10^{10} coulombs over macroscopic volumes, as observed in thundercloud charge separation and large atmospheric discharge systems. A plasma entity could therefore, in principle, achieve significant Lorentz acceleration by maintaining a modest charge imbalance by plasma standards.

Propulsion by Current Filaments

An alternative Lorentz-force mechanism involves the interaction between an electric current and the geomagnetic field, described by:

$$F = I \times L \times B$$

Where the force arises from a current-carrying filament of length L interacting with the magnetic field.

Rearranging the equation to solve for the required current I:

$$I = F / (L \times B)$$

Using the same required force:

$$F = 193 \text{ N}$$

and assuming an internal current filament approximately equal to the entity's body length:

$$L = 96 \text{ m}$$

$$B = 5 \times 10^{-5} \text{ T}$$

Substituting values:

$$I = 193 / (96 \times 5 \times 10^{-5})$$

$$I = 193 / 0.0048$$

$$I \approx 4.0 \times 10^{4} \text{ A}$$

This yields a required current of approximately 40,000 amperes (40 kA)

Thus, a current of approximately 40 kA within the entity would generate the necessary lift or propulsion. This is strikingly similar to the current levels in lightning leaders, which have often produced observations of entities, with similar current levels in sprites, blue jets, and other atmospheric plasma structures, often involving currents in the range 10^3 to 10^5 Amperes. A twelve-metre entity, with a smaller required force, would in turn require proportionally lower currents for the same acceleration.

Implications for Electromagnetic Manoeuvrability

Two main conclusions emerge from these simple Lorentz-force estimates:

1. Earth's magnetic field is far too weak to accelerate a solid, heavy craft of conventional mass using realistic charges or currents.
2. The same geomagnetic field is entirely sufficient to accelerate plasma lifeforms with the masses estimated in chapter one, provided the entity maintains a net charge of a few thousand coulombs, or an internal current filament of tens of kiloamps.

Both conditions are physically typical of natural plasma structures. In such a regime, the entity could use Lorentz forces to accelerate, change direction, maintain lift, or hold position relative to geomagnetic field lines.

This provides a plausible, physics-consistent alternative to anti-gravity technology explanations often invoked in UFO and UAP literature.

Cognitive, Sensory, and Energetic Integration

Both the electroaerodynamic and magnetic fin systems may serve functions beyond locomotion. The same field interactions could underpin perception, cognition, and communication.

Internal cavity resonator arrays nested fractally within the plasma structure may function as computational elements or EM sensors, reacting to changes in surrounding electric and magnetic fields. This would allow the plasma entity to perceive energy sources, magnetic anomalies, or even neurological fields emitted by human observers.

It may also enable field-based interactions with other entities, as well as non-local or synchronised activity patterns. The lateral fins and EAD field surfaces may operate as distributed sensory arrays, not dissimilar to the decentralised neural architectures of cephalopods.

INTERACTION WITH HUMAN NEUROLOGY

Human witnesses of plasma-based phenomena often report unusual physiological effects, such as tingling, dizziness, transient paralysis, or overwhelming emotional sensations. These may result from direct interaction between the entity's field and the human nervous system.

Low-frequency electromagnetic fields, particularly in the E~~e~~xtremely L~~l~~ow F~~f~~requency (ELF) and V~~v~~ery L~~l~~ow F~~f~~requency (VLF) bands, are known to interact with biological systems. When such fields are coherent and resonant with neural signalling pathways, they may induce unusual sensations or modulate cognitive processing.

The appearance of plasma ~~entities~~ beings in domestic spaces, especially near conductive or reflective materials such as mirrors, may lead to enhanced electromagnetic resonance effects. Mirrors may act as charge traps or field concentrators, amplifying ambient disturbances or facilitating standing wave patterns.

These localised electromagnetic environments could create temporary hotspots of neurological influence, leading to visual hallucinations, auditory phenomena, or other altered states of perception.

Energetic and Atmospheric Behaviours

The ability to use ambient atmospheric charge and geomagnetic flux as energy sources or motion vectors makes these entities unique. Their attraction to lightning, aircraft engines, and nuclear sites, which are discussed in later chapters, may relate to their sensitivity to electric or radiative environments.

The same fields that enable propulsion might also allow the modulation of sound resonance with tectonic activity, or even interactions with consciousness via extremely low-frequency coupling. These capabilities suggest a deep energetic and field-based entanglement with Earth's systems.

Conclusion

The evidence assembled in this chapter suggests that the extraordinary manoeuvres associated with UAPs, instantaneous acceleration, impossible aerial manoeuvres, hypersonic movement without shock signatures, silent hovering, trans-medium transitions, and fluctuating visibility, do not require exotic propulsion technologies, anti-gravity devices, or violations of known physics.

Instead, these behaviours emerge naturally when an entity is composed of plasma and interacts with its environment through electromagnetic rather than mechanical means. Electroaerodynamic thrust, Lorentz-force coupling, geomagnetic anchoring, and magnetic self-confinement offer a unified physical framework capable of explaining four of the six observable constraints without invoking engineered craft. A plasma-based lifeform with the ability to modulate charge, current flow, density, and resonant structure could integrate propulsion, perception, communication, and body coherence into a single field-based physiology.

Such an organism would not fly through the air or water in the conventional sense, but would reorganise the electromagnetic space around itself, reducing drag, neutralising inertia, and shaping the medium through which it moves.

Its presence would be a localised restructuring of atmospheric charge, magnetic tension, and refractive index. Behaviour that appears technological from the outside may therefore represent biological function from within.

This model reframes many UAP observations not as demonstrations of advanced machinery, but as manifestations of a plasma ecology, an intelligence possibly evolved within the electromagnetic environment of the Earth.

Rather than representing impossible technology, these entities may be expressions of a deeply integrated, yet hidden biosphere, whose biology is indistinguishable from the fields they inhabit.

The inhabitants of this hidden biosphere may represent a completely different paradigm of life and consciousness, as yet unknown to mankind. Free of our carbon based biology, they would instead inhabit a domain where plasma physics, fractal geometry, and resonance converge, but are they indigenous to Earth?

3 Intelligence Beyond the Veil of Perception

Luminous beings and strange aerial forms have appeared at the edge of human perception throughout history; diversely described as dragons, angels, demons, or UFOs, depending on superstitious, cultural or religious interpretations. In modern times, such phenomena are often dismissed as misidentified aircraft, weather anomalies, or optical illusions.

Yet what if these glowing forms are neither illusions nor machines, what if they are lifeforms? Lifeforms, not in the conventional sense, but something more exotic; beings that are plasma-based, existing partly beyond the familiar fabric of our physical world.

To begin exploring this fascinating possibility, we turn to the concept of interdimensionality, a term that inhabits the borderlands between theoretical physics and metaphysics.

Stripped bare of its more fantastical associations, interdimensionality simply proposes that reality may consist of more than the three dimensions of space and one of time that we ordinarily perceive.

Many of the most advanced physics theories of the modern era, including String theory, M-theory, and holographic dualities, are predicated on the existence of additional spatial dimensions hidden from direct observation. In superstring theory, for example, the universe is thought to possess ten dimensions, nine of space and one of time.

The additional dimensions remain compactified or curled into unimaginably small scales near the Planck length, which is 10^{-35} metres, making them inaccessible to present day experiments. M-theory, a unifying framework that subsumes string theory, suggests an eleven dimensional reality in which matter, and forces exist on multidimensional branes that intersect and interact in higher-dimensional space.

These models imply that a being not confined to our visible dimensions could, under certain conditions, intersect with our reality, projecting only a portion of its higher-dimensional form into our observable three-dimensional space. If life exists in higher dimensions, what medium could anchor it within ours? Plasma, our universe's most abundant and dynamic state of matter, provides an answer.

We established earlier in this book, that as the fourth state of matter, plasma is a partially or fully ionised gas composed of electrons and ions no longer bound into neutral atoms. Unlike solids, liquids, or gases, plasma is inherently responsive to electric and magnetic fields, enabling complex and often self-organising behaviours.

We also discovered that more than ninety-nine per cent of the visible universe is made of plasma, including stars, solar winds, auroras, and interstellar filaments. Unlike the inert solidity of rock or the chaotic instability of gas, plasma is dynamic. It can twist, fold, spiral, and form stable filamentary structures. It can carry and encode information, generate radiation, and respond rapidly to environmental change.

In laboratory experiments, plasma has demonstrated remarkable capabilities. Dusty plasmas, which contain micron or nanometre sized solid particles, can form three-dimensional crystalline lattices known as Coulomb crystals. These ordered arrangements can persist over time, displaying collective dynamics and even forming helical or toroidal structures that resemble the nested symmetries found in biological systems.

Plasma vortices have exhibited soliton-like behaviour, maintaining their shape over long distances and timescales, properties normally associated with coherent systems in nonlinear media. In physics, a *Soliton* is a stable, self-reinforcing, localized wave packet that maintains its shape and constant velocity over long distances, even after collisions with other solitons.

Such characteristics suggest that plasma could serve as a substrate for lifelike behaviour, perhaps even for life itself. In a speculative but increasingly credible field of study, some researchers have proposed that self-organising plasmas in space, or specific atmospheric conditions, might form the framework for exotic, non-carbon-based life.

These structures could not only be coherent, but capable of storing and transmitting information. Given that plasma emits across a broad electromagnetic spectrum, from radio waves to ultraviolet and X-rays, a hypothetical plasma lifeform might encode data, communicate, or sense its surroundings using methods invisible to human perception.

A deeper possibility is that such beings may not exist entirely within our dimension. Their core structure residing instead, in higher-dimensional space, with only a portion protruding into our own and as such, from our perspective, their appearance might change; morphing, shapeshifting, pulsing, or vanish altogether due to their interdimensional nature.

Just as a three-dimensional object moving through a two-dimensional plane would appear to a flat observer as a sequence of changing lines, so a higher-dimensional being might manifest to us as a fluctuating, translucent, luminous form.

The Holographic Principle

There is growing evidence that space and matter are emergent properties of deeper, more fundamental, quantum or informational structures. The holographic principle, derived from black hole thermodynamics and string theory, proposes:

"The information content of all the matter in a three-dimensional region of space can be encoded on a two-dimensional surface surrounding it."

In other words, the maximum information content of a black hole is proportional to the surface area of the event horizon and not the volume inside. Similar to a three-dimensional image recorded onto a two-dimensional holographic plate.

In this framework, our three-dimensional reality might be a projection. An encoded, resonant echo of a deeper-dimensional order. Some researchers have suggested that consciousness and other complex systems could arise from the dynamics of these deeper layers, forming a bridge between information, matter, and perception.

In such a model, a plasma lifeform would not simply be an anomalous phenomenon, but a dimensional lens. A being structured by fields, whose presence modulates or distorts the local spacetime environment. Its body might be composed of oscillating charge filaments and crystalline lattices, responsive both to terrestrial electromagnetic fields and to the geometries of deeper dimensions.

Its mind, if such a term applies, could be distributed across resonant field domains, unfolding and collapsing like a waveform. Such beings might appear in the vicinity of electromagnetic hotspots, vortex events, or altered states of consciousness, perhaps because these are the conditions in which our dimension becomes permeable enough to interface with theirs.

This theory offers a bridge between science and folklore. The Celestial Serpents of Mesoamerican myth, the dragons of Chinese cosmology, the Thunderbirds of the Pacific Northwest, and the Foo fighters and Gremlins of the Second World War, perhaps all represent cultural imprints of the same phenomenon. Plasma beings manifesting under the right environmental and psychological conditions.

Rather than conquest, their primary motivation may be resonance, an alignment between differing layers of reality or consciousness. Their biology would be centred around charge and light instead of chemical compounds, supported by frames composed of geometric lattices of suspended ionised dust particles. Their consciousness, again if such a term applies, may arise from harmonic field interactions that we are only beginning to recognise.

Therefore, we must ask:

"Are they native to a deeper level of reality embedded within the folded architecture of spacetime itself?"

Maybe, a reality from which they materialise when atmosphere, frequency, and perception align. If this model is correct, then the plasma beings we observe and report as UFO's and UAP are not alien in the conventional, extraterrestrial sense.

Although there are alternative psychological, perceptual and cultural interpretations, reports of shapeshifting orbs and luminous entities responding to human observers might reflect entangled perception. A kind of observer-linked effect in which consciousness and plasma fields co-create brief moments of interdimensional visibility.

If so, these lifeforms may be interacting with our reality as a latent presence embedded in the fabric of our world, activated by resonance, attention, or environment. If then, they are not visitors to our reality, then maybe we are witnessing a glimpse of a vast unknown ecology; one of light, vibration and potential, rather than matter.

In this context, is humanity also a participant in that hidden biosphere? An ecology where we may fulfil our potential as a species. If so, the challenge before us is not confrontation or control. Alternatively, we should embrace perception, recognition and understanding.

4 The Plasma Interface of Perception

If these lights in the sky, electromagnetic disturbances, and the fleeting plasma structures that sometimes appear at the edge of vision are not travellers from distant stars, but are actually residents of a realm woven into the very fabric of our reality, then the concept of the Ruliad, introduced by Stephen Wolfram, offers a radical new way of thinking about the presence of energy-based lifeforms and the existence of an hidden biosphere that may coexist with our world.

The Ruliad is not merely a poetic idea. It is conceived as the mathematical limit of all possible computations, the total structure formed by running every conceivable rule on every possible initial condition. In this infinite framework, exist all possible universes, all mathematical truths, all physical laws, and every conceivable path of logic or causality.

Our universe, in this model, is simply one slice through the computational hyper-structure of the Ruliad, defined by those particular physical laws and computational constraints that govern our perception and experience. Nothing lies truly outside the Ruliad. All potential realities exist within it, yet different observers, each defined by their own computational framework, are able to access only certain fragments of its totality.

In Wolfram's interpretation, reality as we perceive it is a consequence of our observer framework. Our brains, bodies, and senses process information in a particular way, and therefore we experience a subset of the Ruliad that matches our biological computation.

Just as Einstein showed that motion and time are relative to an observer's frame of reference, the Ruliad suggests that the laws of physics, causality, and even existence itself are relative to an observer's computational path.

This opens the possibility that other forms of life, plasma-based and electromagnetic, could inhabit slices of the Ruliad that intersect with ours only occasionally, and only under particular energetic or environmental conditions. Such entities would not share our biological substrate of carbon and water.

Instead, they may be built from electromagnetic structures, plasma shells, or field-based consciousness. Their reality would be no less valid than ours but optimised for energy flow and field dynamics rather than the chemical bonding of atoms.

A biosphere of this kind could easily remain invisible to conventional detection. Its inhabitants might function on frequencies, timescales, or spatial geometries beyond our sensory reach. When the two slices overlap, we might observe them as UAP, luminous orbs, or apparitions.

From their perspective, we may appear as slow-moving, chemically bound patterns, perhaps not even recognisable as life. In earlier chapters we have considered the possibility that plasma-based entities might possess intelligence, structure, and intent.

The Ruliad provides a theoretical foundation for this idea. Such beings could exist as stable computational processes, self-organising systems evolving under rule-sets unfamiliar to our physics but entirely valid within the larger Ruliad.

Their appearance to us would be fleeting, shaped by the brief alignment of atmospheric or electromagnetic conditions that allow our perceptual slice to intersect with theirs. Their bodies might consist of oscillating electromagnetic fields. These could be sustained through energy gradients, resonance, and harmonic stabilisation.

Their intelligence could be distributed through field structures, with memory stored in feedback loops or crystalline plasma nodes functioning as a computational centre or brain. In this light, such entities are not supernatural.

They are simply residents of another computational reality within the same overarching structure, which our biology and technology cannot ordinarily access. Seeing the universe through the lens of the Ruliad changes the criteria for what counts as real or alive.

Life, in this broader view, need not rely on DNA, cells, or metabolism. It could emerge wherever patterns are capable of persisting, adapting, and replicating, whether in plasma, dark matter, or pure computation, removing the necessity for alien life to cross interstellar distances.

If all possible rules are contained within the Ruliad, and our slice occasionally intersects with another, contact may already be taking place, constrained only by the limits of shared perception.

The Ruliad does not discard our physical reality; rather, it embeds it within a far larger whole. In this larger structure, the existence of a hidden biosphere of energy-based life is not only plausible but statistically inevitable.

We have never been alone, as we have never been the sole occupants of our reality, just unable to ascertain the true nature of these transitory denizens. Ultimately, the Ruliad offers a framework for reinterpreting unexplained phenomena as glimpses into neighbouring slices of the computational universe, rather than breaches of natural law.

Within this context, the presence of plasma intelligences and other non-biological entities becomes a logical outcome of the totality of possible realities. The greater mystery is not whether they exist, but why we have remained blind to them for so long.

Throughout this chapter, the terms 'slice', 'thread', and 'observer frame' refer to different ways of describing how a system navigates and experiences the Ruliad's total computational structure.

Perception, Consciousness and Communication

We have examined how the Ruliad, the infinite structure of all possible computational rules and realities, might allow for the existence of a hidden biosphere composed of energy and field. One populated by plasma-based intelligences, elusive and ephemeral to our senses, yet possibly as ancient and complex as any biological lineage.

Now we must turn our attention inward. To understand these

beings, we must first examine the limitations and possibilities of our own perception, the nature of consciousness, and the strange boundary where plasma may serve as a bridge between worlds.

Stephen Wolfram posits that what we perceive as the laws of physics are merely a projection of the Ruliad, shaped by our own computational vantage point. We do not witness the totality of what exists; rather, we witness what we are capable of perceiving, what our senses, brains, and cultural paradigms can process and decode.

In this view, every observer is a computational thread, a distinct trajectory through the Ruliad, perceiving only what aligns with its internal structure. This means that what is invisible to us may not be invisible in any absolute sense. Our biology filters the world into patterns that help ensure survival, but it does not guarantee comprehension.

The slice of reality we call the physical world is only a narrow band of a much wider, richer spectrum. Other non-biological intelligences, having evolved in parallel domains, may follow rules adjacent to our own. They might be here, now, moving beside us or in the same space, but in a form our perception cannot detect.

If plasma entities exist within alternate rulesets, branches of the Ruliad with dynamics that differ subtly or drastically from our own, then our first and greatest barrier to direct contact, is perceptual. An alternate, plasma consciousness might experience time as fluid or layered, rather than linear. They might perceive light not as a speed but as a rhythm, or motion as field distortion rather than physical displacement.

What we see as a transient shimmer, an orb of light, or a pulsing glow in the night sky, might be a conversation to them or maybe a direct attempt at communication with us. They may occupy domains of vibration and charge, rather than matter, and so we mistake them for interference, background noise in images, glitches, or fleeting illusions.

Yet it is not they who are insubstantial, but our own ability to register their form that is lacking. Perception, in this framework, is less like a window and more like a filter, one that defines the reality we are capable of witnessing.

If the Ruliad is a universal space of all computation, then consciousness itself may be a computational phenomenon also; a dynamic, self-reflective process that emerges whenever coherence and feedback loops stabilise. In biological life, this coherence is maintained through chemistry, temperature, and tightly regulated systems.

In plasma, it might emerge through stable electromagnetic fields, soliton-like patterns, or scale-free fractal geometries capable of retaining identity amid chaos. Within turbulent environments like the atmosphere or magnetosphere, unstable plasma dissipates rapidly.

If coherent field structures can emerge as vortex rings, filaments, and self-reinforcing waves, they may act as scaffolding for plasma-based cognition. These structures are known to resist disruption, persist through instability and interact with the environment in complex ways.

A plasma entity may thus be defined by the confines of phase and field; a ballet of forces so refined that it becomes self-aware. To persist in such volatile domains, these lifeforms would need to develop strategies to assist with coherence.

Fractal field patterns, resistant to turbulence at all scales, may act as stabilisation. Shifts in density, brightness, and structure could allow an entity to vanish from the visible spectrum, mimic natural forms, or split and reassemble itself within seconds.

What appears to us as multiple beings may be a single entity traversing various excitation states, brightening, folding, and diffusing in ways our perception cannot track. Like all living systems, a plasma-being would be subject to entropy.

In other words, it must acquire energy to maintain its structure, compensate for loss, and to possibly grow, heal or reproduce. While biology relies on sunlight, chemical reactions, or organic digestion, plasma-based life might harvest energy from turbulence itself.

Lightning storms, geomagnetic activity, and swirling atmospheric vortices such as dust devils or willy-willies as they are known in Australia, offer intense, chaotic energy flows. Where we see destruction and chaos, a plasma being might see nourishment, drawn to such phenomena not merely as shelter but as sources of sustenance, using turbulent input as a kind of metabolic scaffold.

Stochastic Resonance

There is even a precedent for this in physics. The phenomenon of stochastic resonance demonstrates; In certain conditions, random noise can enhance rather than degrade the coherence of a system.

As an example, imagine a very faint ticking watch in a quiet room. On its own, the tick is too soft to notice. Now, add just the right, tiny amount of background noise, maybe a gentle hiss, and the ticking actually becomes more audible.

The random fluctuations from the noise occasionally push the ticking sound above your brain's hearing threshold, making it easier to detect. If the noise is too strong, the ticking is drowned out.

If there is no noise, it stays hidden, but at just the right level, the noise helps the weak signal come through. Although counter intuitive, that is stochastic resonance in action.

An entity adapted to this kind of input might not only survive chaos but require it, thriving on disorder in the same way fire needs oxygen and produces movement of the air that we call wind.

If plasma entities are coherent electromagnetic systems, their communication may not be audible or visual in the way ours are. Instead, they might use phase-modulated signals, harmonic resonance, or nonlinear oscillations.

Phase Modulation

These are techniques we already use in advanced telecommunications. A plasma intelligence might maintain an internal standing wave, an oscillating field pattern stabilised over time.

To send a message, it could impose subtle phase shifts onto this pattern, encoding information in the timing rather than the strength of the wave. Another being, tuned to the same base frequency, would detect these phase changes and respond.

Nonlinear Resonance

More advanced still is nonlinear resonance. Here, an entity sends out a wave tuned to a specific harmonic signature. Only a receiver with a matching internal structure would resonate, like a tuning fork, causing another identical fork to vibrate, leaving all else untouched.

This form of communication is private, selective, and invisible to unaligned systems. It could allow plasma intelligences to converse across great distances without betraying their presence to the surrounding environment.

These forms of signalling would be ideally suited to turbulent media, where amplitude based signals are easily lost. In such a framework, communication becomes a matter of aligning patterns, not transmitting vocalisation.

A meeting of fields, rather than a broadcast of words. In chaotic environments, only those configurations that resist degradation will persist. Over time, a kind of evolution must take place, though not biological in the traditional sense.

Plasma lifeforms that cannot maintain coherence will dissipate, so those that adapt by aligning with stable environmental frequencies, healing distortions through phase conjugation, or migrating along coherent channels will endure.

Some may become masters of resonance, using feedback to lock onto environmental rhythms, like birds riding thermals or dolphins surfing waves. Others may fractalise their structure, maintaining form across multiple spatial scales.

Some might learn to reverse distortion, emitting waves that cancel phase shifts introduced by atmospheric interference, preserving identity through the act of interaction.

These strategies are not speculative; they are rooted in principles already observed in plasma physics, nonlinear optics, and chaotic systems. What remains speculative is the possibility that, given enough time and complexity, such strategies may give rise to persistence, perception, intention, and mind.

The answer to how they cross into our awareness, may lie in plasma itself. Unlike solid matter, plasma is reactive, luminous, and shaped by electromagnetic forces. It is the one state of matter most attuned to field-based influence and may act as a kind of visible interface between slices of the Ruliad.

When an orb appears in the sky, or a lightning bolt forks into strange geometries, could it be that a plasma lifeform is briefly intersecting our perceptual thread by adapting its structure? Like ripples on a pond, these flashes may be surface effects of deeper intelligences beneath, manifestations of a computational overlap.

We use technology such as infrared sensors, magnetometers, and satellites to detect what our eyes most often cannot, but it may be that the true frontier is internal awareness, not external instrumentation.

Altered States

Many accounts from individuals who experienced altered states of consciousness, report interactions with visions of light, floating forms, and unknown intelligences. In meditation, trance, near-death experiences (NDE), or through entheogens like DMT and Ayahuasca, many describe contact with luminous entities that shimmer, pulse, or communicate in geometric patterns.

These experiences may not be hallucinations in the traditional sense. Instead, they could represent brief shifts in our computational frame; moments when consciousness aligns with a different slice of the Ruliad. In these altered states, perception may become more fluid, more attuned to harmonic structures and electromagnetic cues.

This does not imply that all such experiences are external encounters, only that some may reflect altered perceptual filtering rather than pure fabrication. When the boundary between self and environment thins, another kind of presence may become perceptible.

If plasma beings communicate through field modulation and harmonic tuning, then perhaps it will not be technology that will allow us to meet them, but resonance of the mind. Under the right conditions, consciousness itself may act as the interface, a soft receiver tuned to another form of order.

If the Ruliad is the totality of all that can exist, then consciousness, life, and meaning are not confined to the familiar. The plasma entities hypothesised here seem to represent an entirely separate thread of evolution.

Their presence may not originate in distant galaxies, but right here, hidden by angle and form. They may not need to travel in the traditional sense of the word, as they are in a sense, already here, woven into our environment, moving at will just outside of our perception.

Stephen Wolfram writes:

"The Ruliad is not just a mathematical object. It is a framework in which everything that can be described, every kind of formal existence, resides."

We live within this framework as does every other intelligence. The challenge is not to build ships to cross the cosmos but to re-tune ourselves to the frequencies, the patterns, and the silences where other minds might already be waiting.

5 Cosmology at the Event Horizon

To the casual observer, the night sky appears as a serene and boundless emptiness; the true depth of distant stars is rarely seen unless far from city lights. Yet for those who watch the heavens, seeking the earliest moments of the creation of our universe across that immense void, the end of their quest may be one step nearer.

The James Webb Space Telescope (JWST) now offers unprecedented glimpses into the earliest eras of cosmic history, and new theories of creation are finding an echo in the data. Recent JWST observations of galaxies formed within a few hundred million years after the Big Bang have yielded a provocative suggestion.

Some early analyses indicate that these young galaxies may share a common rotational orientation across expanses of space so vast that no causal link should be possible. If confirmed, this pattern could represent a primordial imprint in the very fabric of spacetime.

One explanation for such an imprint lies in the idea that our universe was born inside a rotating black hole, known in relativity as a Kerr black hole. The angular momentum of such a progenitor could be inherited by the daughter universe, embedding a preferred spin within its structure. This possibility challenges the principle of perfect isotropy that underpins the standard cosmological model.

In 2010, physicist Nikodem Popławski, working within the framework of Einstein-Cartan theory, proposed that the extreme compression within a black hole need not end in a singularity. Instead, quantum torsion effects could halt the collapse and cause a cosmological rebound, or *bounce*, leading to the formation of a white hole.

From the perspective of the new interior, this would appear as a Big Bang, the unfolding of spacetime from the remnants of a collapse in a greater reality. Physicist Lee Smolin extended this line of thought through his hypothesis of cosmological natural selection. He suggested that black holes may serve as reproductive engines, each one giving rise to a new universe.

Over vast timescales, those universes whose laws of physics favour black hole formation would become more numerous. In this model, the cosmos is not a single creation, but an evolving lineage of realities, shaped by a process akin to biological evolution.

If the early galactic spin alignments suggested by JWST observations are real, they could be the structural memory of a parent universe's rotation, preserved through the transition from black hole to white hole. Such a scenario invites speculation about what might exist beyond the bounds of our spacetime.

If our universe is indeed an enclosed, bounded region within a larger structure, it becomes possible that entities, fields, or intelligences exist outside our domain, capable of interacting with us only at certain interfaces. Such interactions might not involve the travel of matter as we know it, but the crossing of dimensional boundaries, allowing only partial manifestations within our world.

Historical and modern accounts of unexplained luminous phenomena, from floating orbs to intricate, structured light forms, may be reinterpreted in this context. If these are not composed of conventional matter but instead represent organised plasma or field-based configurations, their low density and fleeting visibility could result from the gravitational dissonance between their native environment and ours.

Originating in a realm of extreme curvature, they might only ever project a fragment of themselves into our spacetime. Certain locations on Earth may offer conditions conducive to such manifestations, as we will see in the chapter covering the South Atlantic Anomaly.

The ionosphere, auroral regions, and dusty plasma vortices possess both the energetic and electromagnetic complexity to act as temporary scaffolds for cross-dimensional structures. Observational reports of anomalous activity clustering around storms, volcanic activity, and geomagnetic anomalies lends some weight to this suggestion.

Brane-world cosmology offers another perspective. In this framework, our universe is a three-dimensional membrane embedded within a higher-dimensional bulk. While the fundamental forces of nature remain confined to the brane, gravity may propagate through the bulk, allowing unseen structures in adjacent branes to influence our own.

These models need not be mutually exclusive; each offers a different lens through which boundary conditions, dimensional overlap, and partial manifestation might be understood. This interaction could account for some of the gravitational effects attributed to dark matter.

If life were to evolve in such neighbouring domains, adapted to their distinct physical laws, its projection into our brane could appear ghostlike and incomplete. If additional spatial dimensions were once present but failed to unfold into our observable cosmos, their residual imprint might persist.

In this view, dark matter and dark energy could then be seen as the gravitational echo of geometry folded away from view, curving spacetime without revealing substance and not as hidden particles or exotic matter. Dark energy, by contrast, may reflect the isotropic tension of these compactified realms, projected evenly into our three-dimensional frame.

In this light, the apparent uniformity of dark energy across the universe is not a mystery but the natural consequence of forces diffused through dimensions that did not fully emerge. The holographic principle, first proposed by Gerard 't Hooft and later developed by Leonard Susskind, offers, that all the information contained within a region of space can be described on its boundary surface.

If this principle applies not only to matter, but also to life and consciousness, then plasma-based intelligences could emerge as expressions of information inscribed at the very edge of our universe. In such a model, consciousness itself need not be bound to neural systems alone, but could arise as a standing wave in spacetime, a coherent excitation bridging multiple realities.

Interdimensional plasma intelligences may not stand apart from the universe, but be woven into its geometry, participating in the same deep processes that shape galaxies, fields, and the flow of time. Though speculative, these concepts form a coherent framework for reinterpreting both the physical structure of the cosmos and the unexplained phenomena reported throughout human history.

If our universe is indeed recursive and self-generating, then the luminous entities witnessed in our atmosphere and across our landscapes may be direct expressions of the processes that gave rise to reality itself. The observations of the JWST, coupled with the theoretical work of Popławski and Smolin, point towards a cosmology that unites astrophysics, quantum theory, and the study of consciousness.

In such a view, plasma-based life is not an anomaly but a natural outcome of a multidimensional environment where matter, field, and awareness are inseparably entangled. In this frontier, the mysteries of dark matter, dark energy, quantum entanglement, and the origin of time may prove to be facets of a single geometric truth.

Exploring it will require physicists, biologists, and researchers of many other disciplines, to cooperate in their respective fields, blending measurement with imagination. It is not implausible that our universe lies within the interior of a black hole, nor that we are accompanied by intelligences capable of traversing boundaries we are only beginning to recognise.

What remains is to develop the theoretical language, the observational tools, and the intellectual courage to engage with realities that no longer fit within a single universe.

6 The Liminal Veil

At a distance of nearly 400,000 kilometres from Earth, in the vast gravitational interplay between our planet and its satellite, there exist regions of exquisite balance. Known as Lagrange points, these five zones of gravitational equilibrium allow objects, natural or artificial, to remain relatively stationary with respect to Earth and the Moon.

Among them, L4 and L5 are particularly stable, forming an equilateral triangle with the Earth and Moon and, unlike other positions, they are resistant to perturbation over long timescales. They are quiet sanctuaries in the heavens, shielded from the chaotic flux of solar wind, meteor bombardment and terrestrial weather.

It was amid these zones in 1961 that Kazimierz Kordylewski, a Polish astronomer, first reported faint patches of dust. Using long-exposure photography and meticulous observation, Kordylewski captured the first visual evidence of what are now called the Kordylewski clouds. For decades, their existence was doubted; ethereal collections of fine dust suspended in gravitational stasis at these Lagrange points.

The zones were so faint and the contrast so subtle that many dismissed the clouds as camera artefacts or illusions, until in 2018 researchers at the MTA Wigner Research Centre for Physics in Hungary confirmed Kordylewski's work using linearly polarised light filters and sensitive photometric imaging.

These studies revealed that the dust was real and exhibited polarisation patterns consistent with light scattering through micron-sized particles embedded in a large, diffuse structure. The Kordylewski clouds are most likely governed by Mie scattering, due to the particle size range and the relatively wavelength-neutral, faintly luminous appearance associated with their rare detections.

Mie scattering

Mie scattering occurs where particles are similar in size or larger than the wavelength of the incident light. Rayleigh scattering would only be significant if a large fraction of particles are ultra-fine, which has not yet been confirmed.

Overall, the light scattering behaviour helps explain why the clouds are difficult to detect visually and require specialised observation conditions or polarimetric methods, as used in the 2018 confirmation.

What kind of structures are these clouds? Neither solid, nor gas, these clouds hover at the threshold between the visible and the invisible, like an almost spiritual veil in space. They are not simple accumulations of passive dust; they are dusty plasmas, the fourth, rarely acknowledged state of matter, where charged particles, electrons and fine dust grains interact within a loosely coupled electromagnetic lattice.

As discussed earlier in this book, when observed in laboratory conditions, dusty plasmas exhibit remarkable properties. They self-organise into ordered crystalline structures, maintain long-range coherence and even behave as active media for electromagnetic wave propagation. Such plasmas are dynamic, quasi-living in appearance, shimmering with internal complexity.

The idea that the Kordylewski clouds could be enormous, natural dusty plasmas that stretch tens of thousands of kilometres is not science fiction. It is, at least in part, science fact. What might be their purpose?

To understand the potential of the Kordylewski clouds as more than drifting dust, we must entertain the possibility that nature is capable of spontaneously generating computational architectures, not by utilising silicon, but in dust, charge and plasma.

One of the more provocative theories touching on this notion stems from the field of quantum electronics and, more specifically, the phenomenon known as the *Josephson junction.* This is a device made of two superconducting materials separated by a thin insulating barrier.

When cooled to cryogenic temperatures, these junctions exhibit quantum effects such as tunnelling and phase coherence. They are not merely passive conduits of current; they can store, oscillate and release quantum information. These junctions are sensitive to magnetic fields, can act as resonant cavities and even exhibit what is called macroscopic quantum coherence.

They form the basis for SQUIDs, Superconducting Quantum

Interference Devices, used to detect incredibly faint magnetic fields, such as those produced by the human brain or the Earth's magnetosphere.

The hypothesis, presented in passing in works such as Robert Temple's book, *A New Science of Heaven* is that under the right conditions natural structures could form Josephson-like junctions.

If dusty plasmas in the Kordylewski clouds contain a dense lattice of charged particles, interspersed with zones of differing electrical conductivity and field strength, then a kind of vast, distributed quantum network might emerge.

What follows is not a single claim but a layered hypothesis; physical substrate first, informational capacity second, and cultural interpretation last. This network, though wildly different from silicon chips, could serve similar purposes, including information processing, resonance matching, even a form of predictive computation.

Such a medium could, in principle, record and store the oscillations of every electromagnetic event on Earth. Especially those associated with emotional, spiritual or mass human activity, that which Jung might have called the collective unconscious, or what Eastern traditions refer to as the Akashic records.

If each tiny dust grain serves as a resonator, if phase-locked oscillations propagate through the medium, and if external inputs from Earth's electromagnetic activity are mirrored or absorbed into this matrix, then the Kordylewski clouds may act as vast computational reservoirs. In other words, they may be thinking. To the esoteric traditions of antiquity, there has long existed the idea of a hidden record, an invisible medium that preserves every thought, action and event.

The Akashic Records

The Akashic Records, in Hindu and theosophical systems, are described as an etheric archive of all that has occurred and all that could occur. Not a metaphor, but a literal field of memory woven into the substrate of space itself.

In recent decades, this idea has returned in scientific guise as information fields, zero-point substrates or quantum holographic memory grids. What if the Kordylewski clouds, suspended in the gravitational stillness of Lagrange points, function as a real-world akashic medium?

Their unique position, stable, shielded from terrestrial turbulence and immersed in Earth–Moon electromagnetic resonance, would make them ideal receivers and repositories of subtle informational patterns.

The Earth's Schumann resonance, human biofields, and planetary geomagnetic fluctuations are measurable electromagnetic phenomena, radiating upwards from the Earth and modulating the space around it.

Therefore, dusty plasmas sensitive to charge gradients and capable of long-term oscillatory coherence could theoretically record and retransmit these frequencies. What forms in the clouds may not be static structures but dynamic standing waves, a living archive that reflects the collective state of consciousness on Earth.

In this way, the clouds become an akashic matrix; a layered field of potential, recording not just past events but probabilistic futures, encoded in complex patterns of plasma harmonics. If we accept that plasma-based lifeforms exist and are intelligent, structured and aware, then their relationship with the Kordylewski clouds may be more than incidental.

Plasma-based lifeforms may use the clouds as an interface, like nodes in a larger consciousness network. The complex, crystalline cuboid structures observed within the orb of plasma entities may act as frequency-tuned resonators, allowing them to read from, and write to the Kordylewski matrix.

If each cuboid structure within an entity functions as a quantum cavity resonator, scaled across fractal layers, it could match its internal harmonics to those stored in the dusty plasma fields. This could create a real-time feedback loop, enabling the entity to access stored information and perhaps even influence the environment below.

If that sounds far-fetched, consider the historical echoes: the biblical burning bush that delivered commands to Moses with divine clarity, the pillar of light that led the Israelites by night, the star of Bethlehem that guided wise men across the desert.

Luminous phenomena, sometimes stationary, often moving against expectations, are consistent descriptors of structured plasma phenomena appearing at pivotal moments in human development.

Could these beings have acted as guardians, nudging humanity along a path, not through domination but by benevolent suggestion? This is not a new idea; visionaries, prophets, even artists have long described being pushed or pulled by forces they could not name.

If plasma lifeforms, interfacing with the Kordylewski clouds, have access to a real-time simulation of Earth's collective mind, they may possess the capacity to gently intervene, like stewards of timeline stability. From one perspective, this becomes the architecture of synchronicity and from another, the anatomy of divine intervention.

Across millennia, civilisations have described encounters with luminous entities that emerged from the sky, often accompanied by wisdom, instruction, or warning. These accounts, often dismissed by modern scholarship as myth or metaphor, carry uncanny parallels with what we are now uncovering in plasma physics and consciousness studies.

In Mesoamerican tradition, the deity Quetzalcoatl, the feathered serpent, descended from the heavens bringing knowledge, agriculture and law. It was radiant, serpentine, celestial, and both feared and revered. In ancient China, lóng dragons moved through the clouds, glowing with pearlescent orbs of power.

In Norse mythology, the rainbow bridge Bifröst, connecting Earth to Asgard, may symbolically describe an energetic or plasma conduit. The Vedic vimanas, described as flying palaces, were often radiant, responsive, and governed by thought rather than mechanical propulsion.

These ancient stories often mention light, voice without form, or entities that appear during great upheaval such as war, disaster and revelation. In many of them, the beings do not simply arrive; they manifest, emerging from the sky itself, shimmering at the threshold between worlds.

Could these myths, folklore and ancient texts be early cultural interpretations of plasma entities, temporarily visible as their frequencies intersected with our perceptual domain? Might the "wheels within wheels" of Ezekiel, the fire from heaven that consumed Elijah, or even the angelic visitations described in multiple faiths, be residual memories of electromagnetic intelligences acting through a high-energy interface?

If this is so, perhaps the Kordylewski clouds serve as staging grounds, vibrational nodes where consciousness and computation align. The Kordylewski clouds, viewed through this expanded framework, become something far greater than inert dust. Suspended in the still points of Earth-Moon gravity, bathed in electromagnetic waves from both Earth and the sun.

They are composed of particles capable of forming complex, resonant plasma lattices and as such they are uniquely positioned to serve as consciousness substrates. What arises in these clouds may be an emergent, dynamic cognition; a cloud-based mind.

A vast, subtle and evolving intelligence using processing units that are not silicon chips but instead are dusty plasma cells, capable of organising into Josephson-like junctions. Its memory would not be a hard-drive, more likely it would be a spectrum of harmonics, folded into nested resonant cavities.

Its interface, plasma lifeforms, perhaps born of the clouds themselves and functioning as autonomous nodes. These beings may travel freely between environments; Earth's upper atmosphere, space, the lunar surface, or other gravitational wells where similar dust–plasma interfaces exist.

Their ability to phase into, and out of, visibility, to mimic known forms or subtly influence thought would render them profoundly effective guardians, observers, guides, perhaps even participants in our unfolding narrative. They may not seek recognition, nor wish to alter our trajectory overtly, but when moments of instability arise, when potential futures branch too dangerously, there may be an unrealised intervention.

Maybe a dream, a warning, or a star that leads the way. From all across the globe, stories have echoed for millennia of radiant beings who came from the sky as teachers. Maybe these were not gods in the abstract sense, nor extraterrestrials in anti-gravity craft, but were instead, disembodied voices, orbs of energy, intelligences of light woven from air and fire.

Maybe it was these lifeforms who taught agriculture, medicine, geometry, calendars and ways of being, and then they vanished, or more precisely, faded back into the unseen. Watching from behind the veil. To dismiss these traditions as simple mythology is to overlook the astonishing consistency across cultures that did not have contact with each other.

Aboriginal dreamtime stories speak of Wandjinas, sky beings with radiant heads and enormous black eyes, responsible for shaping the landscape and laws of the land. In India, the Devas were beings of light and vibration who intervened subtly in human affairs. In Central America, the Maya credited glowing entities called Itzamnas, with teaching writing and cosmology.

In almost every case, the method of arrival or communication is non-physical; dreams, meditation, altered states of consciousness, beings walking from the stars or descending on clouds of fire. What if we interpret these stories through a new lens, one grounded in emerging science and our evolving understanding of plasma, consciousness and quantum fields?

Plasma-based lifeforms, whose modus operandi occurs mostly

at the fringe of the visible light spectrum, might appear exactly as just described; dragon or serpentine forms woven from light, accompanied by thunder, harmonics, or a vibrational presence that bypasses language, interfacing directly with human consciousness.

If the Kordylewski clouds and similar plasma-rich zones function as memory matrices or consciousness nodes, then these "sky teachers" might be emissaries of that system; field-based intelligences tasked with preservation, education and course correction.

Not unlike guardian figures in religious or shamanic frameworks, their goal may not be domination or worship, but resonance, helping humanity stay in harmony with a larger cosmic intelligence or plan.

This could explain why teachings across time appear in waves, golden ages, renaissances, as if contact is initiated in critical windows when human perception, geomagnetic conditions, and planetary needs align. These ephemeral lifeforms may represent a co-evolutionary intelligence, one that has long watched, guided and perhaps subtly co-authored our human history.

If they are not native to a distant galaxy, but rather emerge at the atmospheric and orbital thresholds of Earth, their presence may be mediated by interdimensional access. We must wonder how that mechanism of access might operate. One possible gateway is the South Atlantic anomaly (SAA).

The South Atlantic Anomaly (SAA)

In May 2024, Earth was struck by one of the most powerful geomagnetic storms in recent memory, classed as a G5 event.

Triggered by a rapid coronal mass ejection (CME), the event disrupted satellite systems, destabilised global radio communications and restructured the ionosphere.

Nowhere was the storm's influence more pronounced than in the region known as the South Atlantic Anomaly (SAA), a vast depression in Earth's magnetic field, spanning from southern Brazil to the central South Atlantic.

The SAA has long been recognised as a hazard to satellites and astronauts; due to the way it allows high-energy particles from the Van Allen radiation belts to penetrate unusually close to Earth's surface.

What if the SAA represents something more than a radiation risk? What if, under certain conditions, it becomes a porous boundary, a kind of magnetospheric rift through space and dimensional thresholds?

This hypothesis is indeed speculative, but that should not disqualify the inquiry. Many ideas now foundational to modern science were once seen as fringe or fantastical. If we are willing to explore the unknown with disciplined curiosity, the SAA may represent more than a hazardous zone for satellites.

It may also represent a dimensional gateway for an invisible presence. The Earth's magnetic field, generated by the movement of molten iron in the outer core, is not uniformly distributed. Due to asymmetries in the geomagnetic dipole, the inner radiation belt dips closest to Earth over the SAA, resulting in a thinning of the magnetosphere in this region.

During geomagnetic storms, such as the superstorm of 10–11 May 2024, this thinning is further exacerbated, allowing energetic particles to flood into the upper atmosphere. On this occasion, satellites such as the Macau Science Satellite-1 (MSS-1) and Swarm-C recorded severe ionospheric disturbances over the SAA, including enhanced plasma fountains and thermospheric cooling of more than 23 percent in some regions.

Neutron monitors across South America registered Forbush decreases, signalling geomagnetic energy influxes that had complex, long-duration effects. We must also consider how often this type of solar event (G5), has occurred in Earth's history.

Some analysts estimate somewhere between once every twelve and a half years, and forty one years, depending on the specific metrics used. One thing we do know is, there have been many.

If one were to theorise the ideal conditions required for interdimensional plasma-based life to manifest on Earth, then abundant ionisation, reduced magnetic shielding, and increased energetic input fit the conditions. In this context, the May 2024 solar storm would be an ideal candidate. At this point, let's consider that the SAA itself functions as a direct entry point.

During the May 2024 solar storm, its weakened shielding and influx of energetic particles created a zone of increased permeability. Here, dimensional boundaries may have thinned enough to permit access. The idea that the SAA functions as a gateway for plasma-based lifeforms, does not exclude the possibility that there exist other possibilities for dimensional access, or that they even are interdimensional lifeforms.

However, if these beings inhabit a reality in which time is experienced differently, possibly due to a high-gravity environment, the apparent delay between the May storm and the US east coast sightings in November and December need not be a delay at all. From their perspective, the decision to enter and the manifestation here months later, may have directly followed one another.

Gravity and Time Dilation Effects

In general relativity, clocks deeper in a gravitational well tick more slowly when compared to those further out. If an entity can tolerate extreme effective gravity loads, it may also be compatible with environments where gravitational time dilation is significant. T

he time dilation factor is therefore linked to the strength of the gravitational potential in which the clock resides, not to acceleration in motion. Therefore, lifeforms adapted to extreme gravity environments could experience chronology very differently from our own.

During the 2004 USS Nimitz incident, the reported descent of the object was measured to be from 80,000 feet to sea level in 0.78 seconds. This equates to a gravitational force being placed on the object, of approximately 4,050 g.

While this does not suggest that the environment in which such beings normally exist may be comparable to conditions we would only observe in extreme gravitational wells, it does offer that they may be able to readily exist in high gravitational environments such as gravity wells.

In this light, the May 2024 storm and the December sightings might be causally linked in ways that defy our usual sense of chronology.

Eastern ~~United States~~United States Seaboard

Approximately 7,000 kilometres north of the SAA lies the eastern seaboard of the ~~United States~~United States. This region is home to one of the highest densities of nuclear power plants, electromagnetic infrastructure, and urban energy networks in the world.

From Salem and Hope Creek in New Jersey to Peach Bottom, Limerick, Langley Air Force Base (AFB), and the now-decommissioned Indian Point, the corridor is saturated with ionising radiation, microwave emissions, and high-frequency energy fields.

If we posit that plasma-based lifeforms, whether interdimensional or not, require energetic sustenance or are attracted to fields that mirror their internal structure, the ~~UNITED STATES~~United States east coast becomes a logical attractor. In effect, the SAA could act as an entry point, while the ~~UNITED STATES~~United States East Coast is the nearest pasture, rich with high-density energy.

Beginning in mid-November 2024, numerous sightings of unidentified anomalous phenomena (UAP) were reported over New Jersey and adjacent states. Described as silent, hovering, high-altitude formations, these drones often exhibited non-ballistic motion, sharp turns, and prolonged stationary behaviour, traits inconsistent with conventional UAVs.

Some formations were reportedly visible on radar, yet unresponsive to electronic interrogation. While the mainstream narrative focused on surveillance drones, an alternate interpretation must be considered and will be addressed in Part Two of this book in detail. Suffice to say, these "drones", in my opinion, were not mechanical devices in general, but living or sentient plasma forms.

In this model, plasma-beings enter Earth's atmosphere during a moment of geomagnetic vulnerability, as in the May 2024 superstorm. They emerge within the weakened magnetic corridor of the SAA, then follow a natural energy gradient northward, attracted by regions with abundant electromagnetic and radiative output, a veritable smorgasbord of energy replenishment.

Once here, unburdened by the constraints of their native gravity well, they would be free to move at the unbelievable velocities we have often heard reported. Such a migration mirrors biological behaviour, just as whales follow nutrient-rich ocean currents, or birds use the Earth's magnetic field to navigate thousands of kilometres, so too might these unseen intelligences move along magnetospheric paths.

Their brief appearance as structured lights in the night sky, and mimicry of surveillance drones and aircraft, may be the outermost edge of a much larger, electromagnetic ecology. A hidden atmospheric biosphere, only perceptible under precise energetic conditions, and only partially understood by those who witness it.

Scepticism is healthy, yet history is full of cautionary tales where ideas dismissed as pseudoscience were later vindicated by data:

- Continental drift, proposed by Alfred Wegener in 1912, was ridiculed until the 1960s, when tectonic evidence confirmed it.
- Similarly, ball lightning was considered a myth until

multiple military and atmospheric studies documented its existence.

- The notion of black holes, first posited in the eighteenth century and mathematically formalised by Karl Schwarzschild in 1916, remained speculative until the 1990s, when gravitational wave data and high-energy X-ray emissions provided indirect evidence.
- Quantum entanglement, dismissed by Einstein as "*spooky action at a distance*", is now confirmed through violations of Bell's inequalities and has practical applications in quantum communication.

These ideas were all once speculative and their acceptance did not begin with certainty, but with observation, imagination, and patience. Whether these lifeforms are interdimensional or indigenous to Earth, evolving in parallel with biological life, we do not know and can only speculate.

If they have, in fact, surfaced from an interdimensional space, then the theory that they achieved this by utilising the South Atlantic anomaly is a possibility. In this case, the principle remains the same; manifestation requires permeability, resonance, and energetic overlap.

This potential gateway reminds us that the boundary between physics and speculation is porous, and that life may indeed inhabit the unseen dimensions of our world.

Conclusion

The plasma entity hypothesis (PEH), framed within the broader idea of a hidden, possibly interdimensional biosphere, suggests that what we observe and currently name as UFOs, UAPs, drones, or orbs may just be the visible tip of a tremendously larger structure; an ecology of intelligent, coherent, integrated life, electromagnetic in nature, possibly manifesting from an interdimensional gateway under optimal geomagnetic conditions.

If this hypothesis is correct, it implies that the May 2024 storm may not merely be a celestial event, but a cosmic sign, saying "Enter here". One that may have allowed intelligent lifeforms through an opening into our realm for eons past. This 2024 solar event may just have been the most recent arrival, and their appearance on the ~~UNITED STATES~~United States East Coast months later may not have been coincidence.

We are only beginning to question what lives in the spaces between perception and physics, but if we continue to ask questions, fearlessly, imaginatively and with rigour, we may one day come to understand what has always been just out of reach.

7 Spirals of Dust and Light

Throughout these investigations in Perth and the Pilbara region of Western Australia, repeated observations were recorded of faint, plasma-like entities in and around willy-willies, also known as dust devils.

Originally, before the formation of the plasma entity hypothesis (PEH), these observations were purely empirical; only later did theoretical reflection suggest a causal link between the formation of these vortices and the presence of entities in and around these dusty vortices.

Initially, I considered that these entities were simply drawn to the turbulence, perhaps engaging in some kind of plasma entity brawl or mating ritual, or that they were merely enjoying the updraughts. However, a recent insight led me to a striking hypothesis.

What if the entities themselves are exacerbating, or possibly even instigating the formation of willy-willies to loft dust into the atmosphere? Thereby forming a more hospitable medium for them to inhabit and once ionised, heal, reproduce, or reenergise?

Just as plants enrich the atmosphere with oxygen through photosynthesis, plasma entities may engage in a process I have dubbed "dustosynthesis", manipulating thermal convection and lofted dust to generate ionised habitats more suited to their existence.

The term "dustosynthesis" is used here descriptively rather than teleologically, referring to functional environmental modification rather than conscious intent. In this sense, the willy-willy is not simply a weather event, but an act of habitat engineering.

This chapter develops that possibility in greater depth, tying together field observations, atmospheric physics, and plasma science to create a detailed theoretical framework. Willy-willies are small-scale, atmospheric vortices that form in hot, dry regions under calm conditions.

The mechanism of their formation is understood as intense surface heating causing air to rise rapidly through a cooler layer above, with wind shear or other disturbances imparting a spin to the rising column.

The result is a swirling column of air and dust extending tens, to hundreds of metres into the atmosphere. The observed area of manifestation is an operating magnetite processing mine, which covers a huge area.

There, dust is a byproduct of the refining process as well as particulate matter on unmade roads which is constantly ground finer by vehicular traffic. Dust is a major issue there due to the presence of asbestos in the soil alongside the magnetite.

Fig. 2. Pilbara region map adapted from DPIRD Geographic Information Services via Wikimedia Commons (CC-BY-4.0)

This is regularly controlled by water trucks, but often the intense heat almost instantly evaporates any moisture. Willy-willies can manifest anytime the ground is dry, summer or winter.

Summer is exceptionally hot, with temperatures occasionally reaching 50° C (122° F) and often extremely humid. Conversely, winter is mostly dry, much less humid with lower temperatures.

In summer, measured ground temperatures can often be over 60° C (140° F) but is much lower in winter. When willy-willies form, they are much more than an aesthetic curiosity as they are key players in the surface to atmosphere exchange of dust, heat, and incredibly, electric charge.

The Triboelectric Effect

The natural electrical potential of the atmosphere in a desert setting, on a normal day at ground level, is approximately 100 volts per metre. However, this increases dramatically inside the turbulent environment of a willy-willy.

The triboelectric effect ensures that as dust grains collide and rub against each other, they acquire opposite static charges. This leads to vertical charge separation and the incredible electric field strengths inside the column. Grain size and mineral composition, such as quartz or magnetite, also affect the magnitude of charge.

Electric fields within terrestrial willy-willies have been measured exceeding one hundred thousand volts per metre, approaching the breakdown threshold of air under low-pressure conditions. Laboratory simulations confirm that the electrical potential, can in some cases exceed this threshold of approximately one to three million volts per metre, depending on temperature, humidity, altitude, and atmospheric pressure.

At these levels, coronal discharges and weak plasma glow in the blue and ultraviolet ranges under high strength fields. This is consistent with the production of transient plasma formation; however, field evidence remains limited.

Triboelectric charging has also been studied in interstellar dust clouds, and willy-willies have been detected on Mars, where they contribute to planetary electrostatics. Repeated observations in Western Australia, have confirmed the appearance of infrequent, dense body-like structures appearing in and around the swirling dust of willy-willies.

However, partial, amorphous morphologies, luminous and dark irregular orb structures are more common. Frame-by-frame analysis of recordings shows entities apparent coherence in and around the dust plume which suggest more than a random eddy of hot air.

From this pattern of recurrence, one could hypothesise that willy-willies may not merely be attracting these entities, but that their presence may be influential in willy-willy formation as part of a considered interaction with the environment.

The emerging hypothesis is that plasma-based entities actively manipulate the lower atmosphere to exacerbate or generate willy-willies. This process can create a dusty plasma habitat favourable to their existence.

Proposed Mechanism for Propagation of Dusty Plasma

- Entities may establish a strong localised electric field gradient at ground level, destabilising the atmospheric boundary layer and promoting vertical convection.

- Field amplification at the microscale could trigger small updraughts that quickly grow in the unstable thermal environment.

- By shaping the surrounding charge distribution and using their own coherent electromagnetic fields, the entities can bias the rising air to rotate, transforming a chaotic updraught into a stable, coherent vortex.

- Once the vortex reaches sufficient strength, dust and sand are swept up from the surface.

- The lofted dust greatly increases the number of charge carriers in the system.

- Collisions and friction among grains generate further charge separation, and the intense fields inside the column can ionise the dust, creating a dusty plasma.

- This plasma, in turn, provides a medium rich in free electrons and ions through which the entities fields can couple more effectively.

The dusty plasma environment created within a willy-willy may serve as more than a temporary habitat for plasma-based entities. Unlike a purely neutral gas, which would rapidly disperse and lose structural coherence, a dusty plasma maintains the vortex's stability for extended durations.

This enhanced cohesion would not only prolong the life of the column but may also offer plasma lifeforms an electromagnetic framework in the form of a rotating lattice, to which they might anchor themselves, or within which they could maintain a semi-stable energetic state.

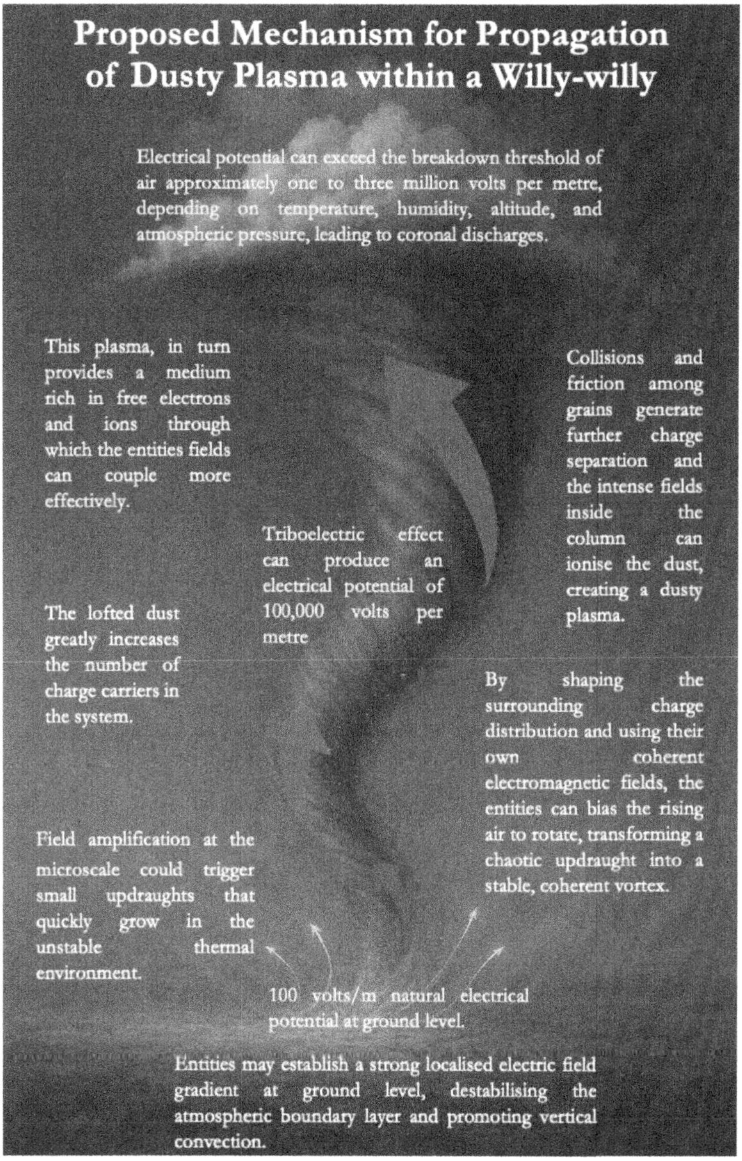

Fig. 3. Conceptual schematic of proposed mechanism for dusty plasma propagation within a willy-willy.

The interactions between the charged dust particles and the ionised gas create a dynamic, self-sustaining system, enabling a level of environmental control that neutral air cannot provide.

Energy Harvesting

Several compelling reasons emerge as to why such entities might initiate and sustain these vortices. Foremost among them is the potential for energy harvesting. Dusty plasmas, by their nature, store a considerable amount of electric potential energy within their charged constituents.

This energy could serve as a direct source of nourishment for plasma-based life, a reservoir they can tap into to maintain form, function, or even growth.

Habitat Formation

Another possibility lies in habitat formation. The plasma-rich environment produced by a willy-willy supports the kind of coherent field structures these beings appear to require. It could act as a kind of self-generated micro-environment, a localised bubble of organised complexity within the otherwise chaotic lower atmosphere.

Within this zone, the entities may more easily maintain their integrity or interact with the world around them in ways we do not yet fully understand.

Reproduction and Communication

It is also conceivable that the vortex plays a role in reproduction. The rotating column may act as a resonant amplifier for electromagnetic emissions, enabling long-distance signalling between entities or potentially triggering as-yet-unknown plasma-based reproductive mechanisms.

The electromagnetic noise of the dusty column, rich with fluctuations and standing waveforms, could also serve as a type of beacon for territory demarcation.

Supporting Evidence

This hypothesis, that plasma entities are seminal in the initiation of willy-willies for purposes of energy acquisition, environmental control, reproduction and communication, is not without some supporting evidence.

Experiments in laboratory settings have shown dusty plasmas forming self-organised, lattice-like, and even helical structures that display complex, dynamic behaviour.

This demonstrates that charged dust in plasma does not behave randomly, but rather forms coherent patterns, some of which bear striking resemblance to the structures recorded in the field. Moreover, field measurements have confirmed that willy-willies can indeed generate ionisation levels and electric field strengths sufficient to support plasma behaviour. ~~Habitat Formation~~

~~Another possibility lies in habitat formation. The plasma-rich environment produced by a willy-willy supports the kind of coherent field structures these beings appear to require. It could act as a kind of self-generated micro-environment, a localised bubble of organised complexity within the otherwise chaotic lower atmosphere. Within this zone, the entities may more easily maintain their integrity or interact with the world around them in ways~~

~~we do not yet fully understand.~~

Laboratory analogues have revealed corona-like discharges and faint glows, often within the blue to violet spectrum, and anomalous radio frequency emissions during strong willy-willy activity.

These phenomena are consistent with weak corona discharges and transient plasma emissions, further reinforcing the idea that something more than swirling dust is at play. These observations may challenge conventional explanations pointing toward deliberate atmospheric manipulation.

Confirmation of Observations

Moving from hypothesis to peer reviewed confirmation will require careful and targeted investigation. Due to the rapid acceleration and speed of movement previously mentioned, one of the most crucial tools will be high-speed, high-frame-rate imaging.

Capturing the earliest moments of vortex formation in fine temporal detail could determine whether the entities are present beforehand or only appear afterwards, thereby helping to establish a causal relationship. However, this task brings its own problems:

Establishing when and where to place recording equipment is crucial, as each willy-willy encounter personally recorded has been opportunistic. Vortices form unannounced and are usually short in duration, a few minutes at best, and often quickly move away from the point of origin.

Spectral analysis is essential. Locating and measuring any specific light emissions from the dust plume would allow researchers to identify characteristic plasma signatures. These signatures would include emission lines that correspond to ionised species such as oxygen, nitrogen, or metallic dust particles, which would confirm the presence of plasma.

Equally important is the deployment of portable electric and magnetic field sensors. Mapping the electromagnetic environment around willy-willies during entity activity could reveal anomalies consistent with directed field manipulation. Measuring and recording differences in ionisation, RF emission, or longevity between entity and non-entity associated willy-willies may offer crucial insights.

Finally, comparative analysis is key. Ideally, willy-willies should also be recorded and studied in areas where entities have not been observed, to serve as control conditions.

While the hypothesis remains unconventional, it is testable. If validated, it would dramatically expand our understanding of both atmospheric physics and the nature of life itself, suggesting that intelligent plasma-based organisms may not just inhabit the Earth's atmosphere but actively engineer it.

Conclusion

The idea that plasma entities may cause or exacerbate willy-willies rather than merely being attracted to them, offers an unprecedented opportunity to link atmospheric physics and the study of unexplained plasma phenomena. It suggests these beings actively shape their environment to create and sustain conditions favourable to their existence, a hallmark of life.

The Pilbara may thus prove to be a crucial natural laboratory, not only for studying willy-willies but also for unveiling the hidden biosphere of plasma-based lifeforms that may have existed virtually unseen for millennia. A signature of life that makes use of dust, charge, and vortex to sustain itself and perhaps others of its kind.

8 Earths Hidden Order

Sacred geometry is the study of forms, patterns and proportions believed to hold symbolic, metaphysical and even cosmological meaning. It rests on the idea that certain shapes and mathematical constants, such as the Golden Ratio, the Fibonacci sequence and Pi, are not merely abstract concepts but fundamental principles woven into the structure of the natural world.

These patterns appear in ancient art, architecture and religious symbolism, but they also emerge in the growth of plants, the form of shells, the symmetry of snowflakes and the spiral arms of galaxies. Sacred geometry is often seen as a bridge between science, spirituality and nature.

It suggests that the same principles shaping a ferns unfolding fronds might also govern the orbits of planets or the arrangement of atoms. Examples in nature are abundant. The chambered nautilus shell grows in a logarithmic spiral, each new section expanding in precise mathematical proportion.

The honeycomb, built by bees, is composed of hexagons,

shapes that fill space with perfect efficiency. Snowflakes form with intricate sixfold symmetry, each one a frozen fractal. The arrangement of leaves and flower petals often follows Fibonacci sequences, a pattern that optimises exposure to sunlight and rainfall.

The Five Platonic Solids

These recurring forms suggest that geometry is not imposed upon nature from the outside but arises from within it, revealing a hidden order that unites living and non-living systems alike.

Among the most revered expressions of sacred geometry are the Platonic solids, five unique, three-dimensional shapes in which all faces, edges and angles are equal:

- Tetrahedron.
- Cube.
- Octahedron.
- Dodecahedron.
- Icosahedron.

Each can be perfectly inscribed within a sphere, its vertices touching the inner surface. This quality implies a kind of internal harmony, a balance between shape and space. To Plato, these solids represented the elemental structure of reality itself, earth, air, fire, water and ether. Each Platonic solid also has a geometric counterpart.

The cube and octahedron are duals, as are the dodecahedron and icosahedron. The tetrahedron is unique in being self-dual. In these relationships, the vertices of one shape correspond to the faces of its counterpart, revealing a symmetry that points to a deeper principle, that opposites such as inner and outer, structure and space, or matter and energy may be connected through form.

For Plato, the perfection of these solids was no accident. They were, in his view, earthly manifestations of higher, unchanging templates that existed beyond the material world. In his Theory of Forms, every object or concept in our reality is an imperfect copy of a perfect, eternal archetype that exists in a more fundamental realm.

The cube carved from stone is a shadow of the perfect, "Form of Cubeness"; the physical dodecahedron is only a reflection of its flawless counterpart. This philosophical perspective connects directly to a well-known story from Plato, the Allegory of the Cave.

The tale has been retold in countless books and classrooms, yet it is worth revisiting here, for it offers a vivid mental image for a central idea in this work; that what we perceive may be only a fraction of a larger reality.

While it is a philosophical metaphor rather than a scientific description, it aligns closely with the projection model proposed here for certain anomalous lifeforms.

Plato's Cave Allegory

Prisoners are chained in a cave, able to see only shadows cast on the wall by objects behind them. Mistaking the shadows for the whole of reality, they remain unaware of the truer, brighter world outside.

In much the same way, the phenomena we perceive, whether geometric perfection in nature or fleeting glimpses of plasma lifeforms, may be mere projections of deeper, multi-dimensional structures, their true form existing beyond the reach of our senses.

This theme is echoed in Plato's Timaeus, where he describes a cosmic craftsman, the Demiurge, who shapes the universe using perfect geometric forms as his tools.

Here, geometry is not decoration but architecture, the blueprint of matter and life itself. In this vision, the cosmos is built from order and symmetry, it's very fabric assembled from timeless mathematical patterns.

Plato also spoke of the World Soul, a unifying principle that connects the realm of forms to the physical cosmos. The World Soul infuses the universe with life, awareness and harmony.

This concept resonates with certain modern views of reality as a living, interconnected system, in which matter, energy, and consciousness are deeply entangled. In the proposed framework, plasma-based intelligences and other non-biological lifeforms could be seen as localised expressions of this greater living structure, cells within a universal organism.

Sacred geometries are not confined to art or mystical tradition. They appear at molecular and atomic scales. The icosahedron and dodecahedron for example, occur in the architecture of certain viruses and in the arrangement of carbon atoms in buckminsterfullerenes, or buckyballs.

DNA, the molecular code of life, follows the Golden Ratio in its helical spiral, and its geometry maps to dodecahedral symmetries. Even at the quantum scale, the shapes of atomic orbitals echo these same geometries, suggesting that the principles governing ancient temples and natural spirals also shape matter at its smallest scales.

Modern cosmology and theoretical physics have uncovered surprising parallels. Some models of the early universe invoke dodecahedral symmetry to account for anomalies in the cosmic microwave background.

Mathematical constructs such as the Amplituhedron and the E8 lattice, used in string theory ~~(Green, Schwarz & Witten, 1987)~~ and quantum field models, describe particle interactions and multidimensional relationships through geometric structures of striking elegance.

Such work implies that geometry may not simply describe the universe but form part of its underlying fabric. In esoteric traditions, sacred geometry has long held spiritual significance. In Hermetic and mystical systems, each Platonic solid is associated with both an element and an aspect of consciousness or a spiritual realm.

The dodecahedron, in particular, was regarded by the Pythagoreans as so sacred that its form was kept secret, thought to embody the etheric or celestial realm beyond the material world. Here, form is seen not merely as shape but as a key to higher knowledge.

Those who have undergone altered states of consciousness, including those induced by entheogenic compounds such as dimethyltryptamine (DMT), often report visions of intricate, shifting geometries, and many describe crystalline forms or rotating solids reminiscent of Platonic shapes.

The mind may either interpret or express the structure of awareness itself in geometric terms, as if consciousness has its own architecture. Carl Jung speculated that archetypes, the deep structures of the psyche, might have geometric or symbolic foundations, aligning with the idea that geometry underlies both thought and reality.

It is mathematically proven that only five perfectly regular Platonic solids can exist in three-dimensional space. This rarity hints at a kind of cosmic selectiveness, where all other forms are variations or derivations of these fundamental shapes.

In this sense, the Platonic solids can be regarded as archetypes of geometric perfection, underlying the manifestation of matter and perhaps even the patterns of experience.

These five forms are also embedded in the *Flower of Life*; an ancient figure composed of overlapping circles. When lines are drawn between specific points, all five Platonic solids emerge from its pattern.

This has led some to regard the Flower of Life as a two-dimensional code for three-dimensional reality, a symbolic blueprint of the universe. Its apparent simplicity conceals a profound depth, echoing the layered elegance of geometry itself.

Through sacred geometry, the universe appears not as a random expanse but as a coherent whole, shaped by underlying principles that are both beautiful and functional.

From the arrangement of petals on a sunflower and architecture of ancient monuments, to the organisation of molecules and the visions of mystics, geometry reveals itself as a unifying thread, connecting nature, consciousness and the cosmos.

9 Crystals and Planetary Form

Beneath our feet lies a planet constructed not only of minerals, but also geometry. At the atomic and mineralogical level, Earth is composed of crystalline frameworks that repeat with precise symmetry, forming the scaffolding of every mountain, plain and seabed.

These crystal systems are not random; they emerge from fundamental physical laws and environmental conditions. Understanding these systems offers more than academic insight; it provides a framework for exploring resonance, energy, consciousness and the nature of UAP-related phenomena.

In this chapter, we delve into the seven crystal systems, how they manifest in Earth's geology, and why their geometries might be important to plasma-based lifeforms and energetic anomalies explored in this book.

We will also examine the distribution of crystal systems across geological regions, especially in Western Australia. We will also uncover the geological systems of the Moon, to provide a reference point for the chapter covering Transient Lunar Phenomena.

Every crystalline material on Earth conforms to one of seven fundamental symmetry systems, based on how atoms or ions are arranged in three-dimensional space. These systems define the shape, symmetry and energy behaviour of the resulting mineral.

The Seven Crystal Symmetry Systems

The cubic or isometric system is the most symmetrical of all, with three axes of equal length intersecting at right angles. It produces perfect cubes and highly symmetrical structures, often found in minerals like magnetite, pyrite, halite and diamonds.

These minerals frequently exhibit strong magnetic, electrical or optical properties. The tetragonal system is similar to the cubic but has one axis longer or shorter than the other two, while all angles remain at 90 degrees, in effect a rectangular prism with a square base.

This system gives rise to elongated or compressed forms and includes minerals like rutile and zircon, which can exhibit strong optical behaviours and are important in piezoelectric and high-temperature applications.

The orthorhombic system has three unequal axes, all intersecting at ninety degrees. It creates prism-like or rectangular crystal shapes and includes common minerals such as goethite, sulphur and barite. These often occur in iron-rich soils and banded formations.

The monoclinic system also has three unequal axes, but only

two angles are at ninety degrees, with the third skewed. This creates a more tilted rectangular structure. Monoclinic minerals include gypsum, muscovite and monazite, many of which form in metamorphic or hydrothermal environments and sometimes include radioactive elements.

The triclinic system is the least symmetrical, with all three axes unequal and none of the angles at ninety degrees. This gives rise to highly skewed and distorted crystals, such as those found in plagioclase feldspar and kaolinite. These are common in granites and clay-rich environments.

The hexagonal system includes three equal axes at 120 degrees and a fourth vertical axis of different length. It forms six-sided prisms and includes minerals like quartz and beryl, which are important for their piezoelectric properties and frequency stability.

The trigonal system, often grouped with hexagonal, features three equal-length axes intersecting at non-90-degree angles. This results in rhombohedral or tilted-cube shapes. Hematite, calcite and corundum all form within this system. These minerals often display substantial optical and magnetic behaviours.

These seven symmetry systems govern how minerals interact with light, magnetism, vibration and possibly consciousness. In our plasma-entity model, these geometries may serve as resonant templates or morphic fields for energy-based lifeforms or high-frequency structures.

Different geological environments favour the formation of specific crystal systems. In Western Australia, the Pilbara region, and particularly the Hamersley Province, is dominated by cubic and trigonal systems. It is rich in magnetite and hematite, minerals whose magnetic properties may be relevant to plasma entity formation or behaviour.

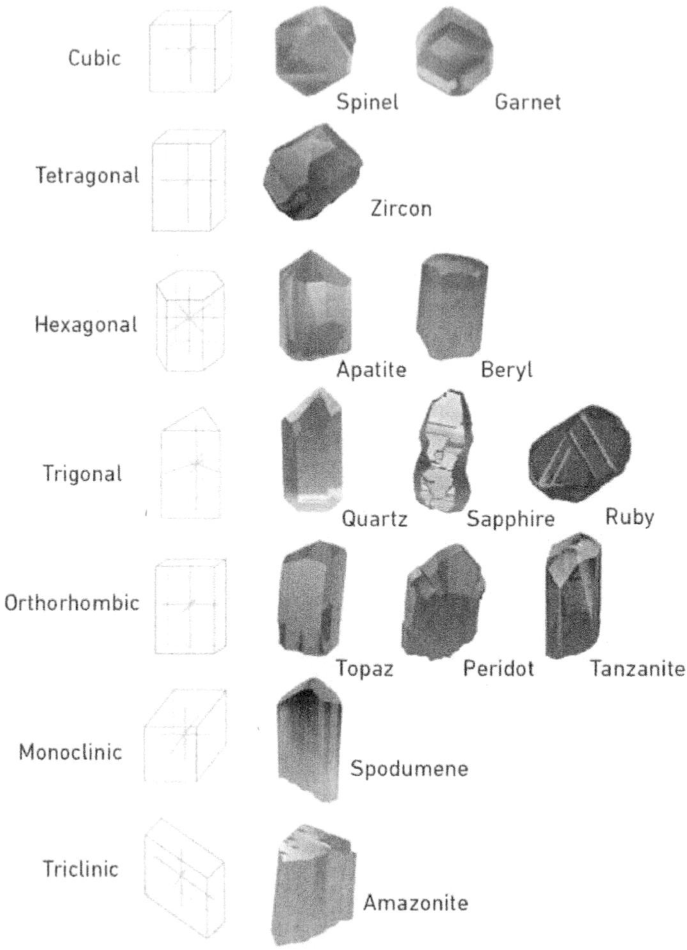

Fig. 4. The seven crystal Symmetry Systems.

The Yilgarn Craton, encompassing much of south-west Western Australia including the Perth region, presents a complex mixture of orthorhombic, triclinic and monoclinic systems. Minerals such as goethite, feldspars and kaolinite dominate the landscape. While these minerals are not strongly magnetic, they offer structural and compositional diversity.

In the Kimberley and northern Western Australia, hexagonal systems are prevalent, especially in quartz-rich zones. The presence of piezoelectric minerals in these areas may be significant for energetic anomalies and transient plasma formations.

The Mid-West and Goldfields regions host a broad range of mineral systems, with monoclinic and orthorhombic structures being common. These areas are rich in rare-earth and thorium-bearing minerals such as monazite, which may have implications for natural radiation and energy interactions.

Lateritic zones such as the Darling Range tend to be poorly crystalline or amorphous. These areas are dominated by altered forms of goethite and bauxite, which may create incoherent or unstable field conditions for plasma phenomena.

Some minerals, especially those containing radioactive elements like uranium and thorium, begin with well-defined crystal systems but degrade into amorphous structures over time due to internal radiation damage. This process, known as metamictisation, can disrupt the electromagnetic properties of the mineral and alter its interaction with energy fields.

Zircon, which starts in the tetragonal system, may become amorphous through long-term radiation exposure. Monazite, a monoclinic phosphate mineral rich in thorium, also frequently displays metamict textures. These transformations may contribute to unstable electromagnetic zones.

A central hypothesis of this book is that certain crystalline geometries act as templates or attractors for plasma-based entities. These structures may provide natural cavity resonators for electromagnetic stability, enable geometric coherence of charge and field, and serve as biological or consciousness scaffolds for field-based life.

In this view, Earth is more than a geological body. It is a frequency lattice, encoded with crystalline geometries that may guide, nourish and encompass non-visible entities operating in plasma and light. This understanding will be extended in the chapter which focuses on Transient Lunar Phenomena (TLP).

There, we will examine how similar crystal systems may exist within the Moon's geology, and how their structure correlates with locations of reported flashes, glows and anomalous lights. The terrestrial crystal system geometries observed, may find their analogue in lunar lattice forms, suggesting a continuity of crystalline influence across celestial environments. This lattice may not end on Earth.

10 Geology and Plasma Lifeforms

Across the world, certain landscapes have become associated with luminous anomalies that defy reduction to meteorology or misperception. The Pilbara of Western Australia, the Hessdalen valley of Norway, and the Appalachian ranges of the eastern ~~United States~~United States each possess a long and consistent history of floating lights, structured orbs, radiant shapes, and other phenomena suggestive of coherent atmospheric plasmas.

While such accounts are often framed as folklore, optical illusion, or psychological projection, their geographical clustering raises a deeper and more compelling question: what is it about these particular terrains that sustains the persistent appearance of light-like intelligences? If plasma-based lifeforms exist, then their distribution should not be random.

They would favour locations rich in the electrical, mineralogical, and atmospheric ingredients necessary for plasma generation and stability. These regions, with their unique geological histories, crystalline frameworks, aquifer systems, and tectonic dynamics, may form the natural "ecologies" in which plasma entities emerge, organise, and perhaps even evolve. Their behaviour, morphology, and persistence may reflect the deep fabric of the Earth beneath them.

Understanding these terrains therefore becomes an act of ecological interpretation, revealing a hidden biosphere whose constituents are built not from carbon and water, but from charge, structure, and resonance.

The Pilbara Craton

The Pilbara region of Western Australia is one of the oldest surviving fragments of continental crust on Earth, a place where 3.5 billion year-old rocks lie exposed to the atmosphere. Its mineral fabric is dominated by iron ore, magnetite, hematite, and ancient banded iron formations whose crystalline geometries are overwhelmingly cubic or orthorhombic.

The Pilbara can therefore be seen not merely as a landscape of minerals, but as a vast, naturally occurring lattice of geometric order, capable of interacting with electromagnetic fields in distinctive and powerful ways. This region is riddled with faults and fracture zones. Although it is not tectonically dramatic, slow strain accumulates within its aged crust, producing micro-seismic activity and sustained mechanical stress.

Where this strain intersects with quartz-bearing strata and deep aquifers, a chain of electrochemical processes begins. Water flowing through such fault zones becomes ionised; quartz under pressure releases electrical charge; and mineralised layers develop strong localised fields. These electrical conditions frequently give rise to corona discharge and atmospheric ionisation.

It is in this setting that anomalous lights are reported. Many accounts describe glowing orbs drifting above the ground, but perhaps most intriguing are the cuboid or rectangular lights observed at distance or captured in digital recordings. These sharp-edged forms are anomalous within atmospheric optics but entirely consistent with the cubic symmetry of iron-dominated mineral systems.

Plasma, when interacting with strong crystalline-imposed magnetic fields, can adopt shapes that reflect the resonance template of the underlying geology. Ferrofluids form spikes beneath magnets; plasma filaments align within toroidal fields; iron-rich terrains may imprint a similar geometric influence upon atmospheric plasmas. The Pilbara's mineralogical rigidity may therefore be shaping the very bodies of the entities seen within its deserts.

My own observations in the region repeatedly highlight this geometric coherence. During recordings near willy-willies, faint, pulsing forms emerge that possess internal structure inconsistent with optical artefact. Frame-by-frame analysis reveals tessellated surfaces and luminous edges reminiscent of the cuboid resonators proposed elsewhere in this book. The conjunction of geology and plasma here is not coincidental, but fundamental.

Hessdalen Valley

Where the Pilbara is ancient, arid, and iron-bound, Hessdalen occupies a very different geological and atmospheric niche. Situated in central Norway, this narrow valley has been synonymous with mysterious lights since the mid-twentieth century, with thousands of sightings and multiple scientific investigations. Unlike the Pilbara's desert, Hessdalen is a landscape shaped by rivers, moraines, glacial tills, and sedimentary basins. Yet beneath this pastoral surface lies a complex geological engine: sulphide ores containing zinc and copper, quartz-bearing formations, and mineralised fracture networks that interact intimately with circulating groundwater.

The aquifers of Hessdalen play a crucial role. Water flowing through sulphide deposits can trigger electrochemical reactions capable of generating strong electric potentials. When these electrical charges accumulate and discharge into the atmosphere, ionisation occurs. This process, occurring within a valley prone to temperature inversions, creates a partially enclosed chamber in which plasma can persist far longer than it would in open air. The valley acts as a natural electromagnetic vessel.

The structure of the Hessdalen lights differs markedly from those of the Pilbara. Rather than cuboid or block-like forms, Hessdalen phenomena are frequently elongated, prismatic, or spindle-shaped. They divide, merge, or oscillate in colour, behaviour that mirrors transitions in laboratory plasmas undergoing shifts in density or charge distribution. These shapes may reflect the orthorhombic and tetragonal mineral systems of the valley's ores, offering a geological explanation for the entities' distinctive geometry.

Scientific observations have confirmed their objective physicality. Radar has detected structured returns; optical spectra reveal temperature ranges consistent with plasma; magnetometers record disturbances concurrent with sightings. The Hessdalen lights are therefore neither illusion nor anecdote. They are physical, dynamic, and sometimes astonishingly stable plasma structures. Their persistence indicates a system in which geological, atmospheric, and electromagnetic factors interlock with rare precision. If plasma entities require long-lived charge environments to sustain coherent form, Hessdalen provides it.

The Appalachian Corridor

The Appalachian Mountains, stretching along the eastern ~~United States~~United States, represent one of the oldest mountain chains in the world. Their rocks, granites, schists, gneisses, and vast networks of quartz veins, have undergone hundreds of millions of years of folding, metamorphism, and weathering. Beneath the mountains, extensive aquifer systems circulate through layers of fractured quartz-rich rock, creating a natural foundation for sustained electrical activity.

Reports of luminous phenomena in the region date back centuries. Cherokee accounts speak of hovering lights and glowing orbs along ridges, long before colonisation. Nineteenth-century observers documented lights that drifted across Brown Mountain and other Appalachian slopes, sometimes appearing to behave with curiosity or intention.

These accounts reveal a continuity of observation that spans cultures and centuries. Quartz is central to the Appalachian plasma ecology. When quartz is stressed or fractured, it produces electrical charge and, under certain conditions, visible luminescence.

Laboratory studies show that quartz subjected to strain can emit plasma bursts capable of forming small, luminous structures. Unlike the Pilbara or Hessdalen, where mineral geometry shapes plasma configuration, the Appalachians offer a more uniformly quartz-dominated environment.

As a result, the lights here tend to manifest as spherical, prismatic, or amorphous forms, reflecting the less rigid symmetry of quartz compared with iron ores or sulphide systems.

Witnesses describe lights capable of hovering motionless, accelerating instantaneously, or following ridgelines. Their behaviour suggests interaction with geomagnetic gradients rather than inertia. This movement is consistent with plasma structures riding field lines or responding to charge differentials within the crust.

While these orbs rarely exhibit the geometric precision of Pilbara lights or the prismatic forms of Hessdalen, they display a dynamic and adaptive behaviour that indicates more than simple discharge phenomena.

Earth's Electrical System as a Living Framework

All three regions share a fundamental characteristic: they are situated within highly active geoelectrical environments. The Earth's crust is permeated by telluric currents, natural electrical flows influenced by solar wind, geomagnetic storms, conductivity variations, and tidal forces. These currents vary seasonally and respond to cosmic conditions.

During periods of heightened solar activity, many regions with anomalous lights report increases in both frequency and intensity of sightings. Telluric currents are strongest where conductive minerals are present, where water provides ionic pathways, and where faults create electrical discontinuities. Each of the three regions examined in this chapter meets these requirements with unusual intensity.

Their rocks conduct, their aquifers circulate; their faults accumulate charge; their atmospheres provide the medium in which electrical expression becomes visible as plasma. In this context, plasma forms appear not as accidents but as emergent expressions of the Earth's electrical life. Charge becomes structure; structure becomes motion; motion becomes behaviour.

If plasma entities are coherent and self-stabilising, they may rely upon these geoelectric systems in much the same way biological organisms rely upon oxygen, nutrients, and habitat. The Earth becomes a living field of potential energy, shaping and sustaining a hidden biosphere through charge instead of chemistry.

If plasma entities depend upon the Earth's deep electrical cycles for stability and sustenance, then their distribution should mirror that of other ecological niches. Geologists already map the planet by mineral province, crustal stress, and groundwater dynamics; these same maps, reinterpreted through the PEH framework, become potential habitat charts for plasma organisms.

The Pilbara represents a high-iron, high-resonance environment; Hessdalen valley offers sulphide-driven electrochemical gradients, and the Appalachians provide sustained quartz-based charge release. In this sense, plasma life may not be rare but merely regional, its visibility controlled by the geological scaffolding beneath it.

A hidden ecology may therefore span the Earth, like an electromagnetic mycelial network, branching, pulsing, and evolving in ways that only become visible under the right atmospheric and electrical conditions

Piezoelectric Landscapes

Where quartz-bearing rocks are subjected to strain, the piezoelectric effect triggers the release of electrical charge. When water is present within fracture zones, the effect becomes amplified, producing powerful and sustained electrical gradients. These gradients can ionise air, creating the initial conditions for plasma formation.

The Pilbara, Hessdalen, and Appalachians all contain dense quartz systems intersected by faults and aquifers, making them ideal candidates for piezoelectrically driven luminosity. Laboratory studies have shown that small-scale piezoelectric emissions can generate miniature plasma domains that persist for fractions of a second.

Nature, however, provides a far larger canvas. In regions where strain accumulates slowly and continuously, these electrical conditions can be sustained for minutes or hours, potentially enabling the emergence of more complex or self-organising plasma structures. Many luminous anomalies in these regions coincide with periods of drought, groundwater fluctuation, or tectonic stress, conditions known to amplify piezoelectric activity.

Earthquake Lights

Earthquake lights, verified worldwide in seismically active regions, offer a natural analogue for the kind of plasma phenomena described here. These lights are known to precede or accompany tectonic rupture and appear as structured, sometimes colour-changing luminous forms.

They often move laterally, hover, or exhibit behaviour that seems responsive to the surrounding environment. Their existence demonstrates that geological strain alone is capable of producing plasma structures with surprising degrees of stability and organisation. Earthquake lights, however, are short-lived, arising only during intense tectonic stress.

The three regions examined in this chapter appear to host more persistent forms of strain, producing sustained electrical environments rather than explosive discharges. This continuity of energy supply may allow for the emergence of stable plasma ecosystems rather than the transient flashes seen during earthquakes.

Geology as a Plasma Ecology

Taken together, the Pilbara, Hessdalen, and Appalachian regions reveal a consistent pattern: plasma-like phenomena arise where mineral geometry, electrical activity, atmospheric conditions, and groundwater flow coincide in particular ways.

The Earth's crust, through its crystalline structures, its internal stresses, and its electrical currents, may be shaping the structural geometry and behaviour of plasma entities. None of these mechanisms require cognition or intent; yet they establish the physical conditions under which such properties, if they exist, could emerge.

If plasma life exists, then its anatomy would not be defined by cells or organ systems, but by the properties of the landscape itself. The geometry of local crystal systems would shape the forms that atmospheric plasmas naturally adopt. Charge distributions along faults and mineral seams would influence their stability and behaviour.

Aquifer pathways would modulate conductivity and feed sustained electrical gradients into the atmosphere. Regional geoelectric activity would determine where these forms can persist and flourish. Even the coupling between crustal electrical systems and the broader atmosphere would act as a kind of metabolic environment, offering energy, structure and mobility.

In this sense, geology does not merely host these entities; it participates in their formation. Rock becomes scaffold, natural energy becomes sustenance, and the atmosphere becomes the space in which life-like plasma forms manifest. Regions with the right mineralogical, hydrological, and electrical properties become ecological niches for plasma entities, much as forests or coral reefs serve as habitats for biological organisms.

This interpretation suggests that the Earth may contain many such plasma ecosystems, some known through folklore, others yet to be recognised. Geological maps become potential habitat maps; fault lines may serve as migration corridors; aquifer networks may act as energetic arteries; crystal systems may determine morphological variation within plasma populations.

Conclusion

The Pilbara, Hessdalen, and Appalachian regions reveal a remarkable convergence between geology and plasma phenomena. Their mineral compositions impose geometric influence; their aquifers supply charge mobility; their faults generate electrical stress; their atmospheres provide the medium in which luminous forms appear.

If plasma entities exist, these regions are natural haunts, stable, energy-rich, and structurally conducive to the emergence and persistence of coherent plasma life. Seen in this light, the Earth becomes more than a stage upon which plasma anomalies appear. It becomes the very womb in which plasma intelligence may arise.

The mineralogical skeleton of the planet, shaped across billions of years, may have given rise to a parallel biosphere expressed not in organic tissue but in charge and resonance. These entities, luminous and enigmatic, may be neither spirits nor artefacts, but inhabitants of an electromagnetic ecology as old as the crust itself.

11 The Lunar Mirage

Transient Lunar Phenomena

Are anomalous phenomena limited to our terrestrial home? The same interplay of crystalline substrates, charged dust, and external energy inputs that may shape plasma activity here on Earth may also exist on the Moon.

If intelligent plasma lifeforms can emerge within the dynamic, water-rich and electrically complex environment of our own planet, then is the austere exosphere of Earth's satellite also suited to their appearance?

For centuries, observers of the Moon have reported flashes, glows, obscuration and other strange, momentary effects upon its surface. These events, collectively referred to as Transient Lunar Phenomena (TLP), have been documented by both amateur and professional astronomers since at least the 6th century, with a surge in reports during the telescopic age.

Despite frequent documentation, TLPs remain elusive, unexplained, and largely excluded from mainstream lunar science. The Moon offers us a distant stage upon which plasma phenomena may reveal themselves. This chapter visits TLP through the lens of plasma physics and planetary geology, building on our earlier exploration of crystal systems.

It is theorised that many TLPs may not be caused by impact events, outgassing, or observational artefacts alone, but instead could represent the visible activity of atmospheric plasma-like beings within the unique conditions of the Moon's exosphere.

Drawing on parallels from terrestrial phenomena such as willy-willies in the Australian outback and beyond, we will consider how electromagnetic interactions, dust, and crystalline substrate materials might support self-organising plasma structures in the lunar environment.

Transient lunar phenomena are defined as temporary changes in the appearance or luminosity of the lunar surface. Reports typically involve sudden bright flashes, coloured glows, temporary obscuration, or unusual mists or fogs that can last from seconds to hours.

These events have been recorded in notable locations such as Aristarchus Crater, Plato, Copernicus, and the vicinity of the lunar terminator. While some TLP reports may be explained by outgassing, micrometeorite impacts, or viewing conditions, the consistency of certain hotspots and the spectral features observed suggest that a more coherent physical explanation may be warranted.

High-energy events that produce ionisation or electromagnetic radiation are already known to occur through solar wind interactions, micrometeorite bombardment, and electrostatic levitation of lunar dust. In several reports, coloured glows, particularly red, green, and violet, have been noted, hinting at ionisation and plasma excitation processes.

On Earth, plasma-like atmospheric entities have been observed in the form of glowing orbs, sky flashes, and moving lights associated with phenomena such as ball lightning, will-o'-the-wisps, and willy-willies. As described in the chapter on willy-willies, in arid regions such as the Pilbara these pillars of swirling dust and air often exhibit complex interactions with airborne dust and charged particles.

Dusty plasmas are known in laboratory settings to form crystalline or lattice-like arrangements, where micron-sized charged particles suspended in a plasma arrange themselves into stable geometric forms. These dusty plasma crystals exhibit behaviours that resemble both solid-state and fluid dynamics, including wave propagation, oscillation, and self-healing structures.

Under the right atmospheric and electromagnetic conditions, such as those near the Earth's surface or in controlled environments, these lattice formations can remain coherent for extended durations. The presence of similar electrostatic and dusty conditions on the Moon suggests that comparable phenomena could occur transiently in a lunar environment.

The lunar surface is constantly bombarded by solar UV and solar wind particles, causing photoelectric charging. Dust grains on the surface and in the near lunar environment can become electrostatically repelled and lofted into space. These conditions create a natural environment for the formation of plasma sheaths and transient structures.

Regions of frequent TLP reports often correspond with geologically distinct areas, particularly impact craters and volcanic features rich in silicates, oxides, and fractured substrates. Aristarchus Crater, for example, is known to be compositionally rich in silicate glass and possibly residual volcanic gases. Plato and Tycho craters display high-albedo materials and brecciated surfaces, while areas around the lunar maria show basaltic flows and mineralogical layering.

When we apply the framework of crystal systems outlined in the previous chapter, we find a match between TLP hotspots and regions where minerals such as plagioclase (triclinic), ilmenite (trigonal), and pyroxene (monoclinic) are abundant. These minerals could provide the structural templates or surface interactions needed to support momentary plasma crystal formation.

Just as dusty plasma lattices on Earth can form in the presence of electromagnetic fields and particulate media, the Moon may host ephemeral analogues, shaped and bounded by its own mineralogical fingerprint. It is plausible to consider that TLPs are not merely optical illusions or singular impact events but the emergent visibility of plasma-based entities or coherent dust-lattice formations.

These may arise in response to solar activity, charged dust flow, or even seismic disturbances within the lunar crust. The crystalline geometry of the lunar substrate may shape these plasma structures in ways analogous to the templating seen in terrestrial dusty plasma experiments.

By synthesising lunar geology, observed TLP patterns, and known plasma behaviours, we can begin to form a new hypothesis; that the Moon occasionally hosts transient plasma structures formed by the interaction of electrostatic dust, solar energy, and crystalline surfaces.

These structures may be visible as flashes, glows, or shifting obscuration. Their form may mimic geometric constructs, particularly diamond, rhombohedral, or prismatic shapes, reflecting either the resonant conditions of the local field or the underlying crystal lattice system of the terrain.

This hypothesis draws from both observational evidence and theoretical modelling. It offers a non-anthropocentric interpretation of TLP as a natural consequence of lunar environmental dynamics, while also hinting at the presence of complex, possibly life-like behaviours within non-biological systems.

If plasma entities exist, and if their emergence is influenced by crystal-lattice structures, then the Moon, long thought to be geologically inert and airless, may prove to be a more dynamic and structured environment than previously believed.

12 Symbiosis and Humanity's Hidden Light

The Human Aura

The aura may be understood as the marriage of two subtle forces within the body. On one hand, countless biochemical reactions release photons, faint pulses of light born of oxidative metabolism. On the other, streams of ions course through our nerves, muscles, and membranes, generating fluctuating electromagnetic fields.

Together these emissions do not remain trapped inside us but extend outward, forming a sheath of charge and light beyond the physical body. It is this plasma-like, electromagnetic envelope, woven from photons and ionic currents, that seers have long described as the aura.

Kirlian Photography

A process called Kirlian photography captures similar corona on film, where faint light radiates from living tissue. The subject, when photographed in this way is not seen merely as a physical form, but is wrapped in a subtle plasma-like body which can be photographed. An emanation of electromagnetic light and energy.

However, the process of Kirlian photography is achieved by passing a voltage through the photographic subject, causing artificial corona in the air around the object.

PRIMORDIAL POWER GENERATORS

The causal mechanism of the enigmatic phenomenon of the aura is much stranger and involves a long standing relationship between two organisms. Deep within each of our cells, lie small, bean-shaped structures called mitochondria. At first glance they appear unremarkable, yet they are the silent powerhouses of life.

Each one is wrapped in a double membrane, the inner layer folded and curved into ridges known as cristae. These folds create space for an extraordinary process that sustains every movement, every thought, every breath we take. Mitochondria generate adenosine triphosphate (ATP), the universal energy currency of biology.

They achieve this by passing electrons along a chain of proteins embedded in their inner membrane. As the electrons move, protons are pumped across the membrane, building a reservoir of charge. When the protons flow back, they drive a rotary molecular machine that produces ATP. It is a process as elegant as it is essential, a microscopic turbine that spins endlessly within us.

This flow of ions is more than chemistry. It is electricity at the scale of life, a ceaseless current that surges through the inner walls of the mitochondria. In the act of generating energy, they also release light, faint pulses known as biophotons, emerging as a by-product of oxidative reactions. In this sense, the mitochondria can be seen as bio-plasma reactors.

They are tiny engines coupling streams of ions with flickers of light, creating both the charge and the photons that shape the body's electromagnetic field. Multiply this by the trillions of mitochondria within us and the result is a radiant system, a body that officially glows imperceptibly, a sheath of charge and light extending beyond the skin.

If we call this field the aura, then mitochondria are its foundation stones. They not only provide the fuel for muscle and brain but also for the subtle envelope of plasma-like energy that surrounds us.

In using the term symbiosis here, I am not suggesting a one-to-one biological equivalence with parasites, commensals, or mutualists as classified in ecology. Instead, I am proposing an energetic form of symbiosis, in which plasma entities and humans share and shape a common informational and electromagnetic environment.

We supply attention, narrative, emotion, and in some cases structured expectation. They, in turn, may supply patterning, guidance, disruption, or catalytic experiences that alter the trajectory of individual lives and cultures. The exchange is rarely overt or symmetrical, but it may nevertheless constitute a long-running relationship in which both parties are changed.

Without that moment of symbiosis there would be no animals, no plants, and no humans. We are alive because of a merger so profound that it reshaped the course of evolution of life on Earth. The mitochondria remain as our silent companions, generating energy in every heartbeat and breath. They are not us and yet they are us, bacterial symbionts so deeply integrated that our existence depends upon them.

It is a reminder that life is not an isolated affair but an ecology, a constant negotiation between organisms, fields, and energy. If such symbioses could occur at the cellular level, why not at higher levels of organisation? Just as bacteria became our internal partners, could plasma lifeforms act as external cohorts?

Perception of Energy Fields

What mystics have described for centuries as a glowing energy field, or the subtle body, may have its roots in the ceaseless labour of these ancient symbionts. These photons can be detected with sensitive instruments, confirming that the living body is never in darkness.

Many people report the ability to see and sense this subtle energy field. Indeed, I have experienced this many times. I discovered that when observing a person, directly focusing on their physical body is ineffective in exposing the aura. However, using a more indirect view, whereby the subject is observed in the periphery of the field of view, the white light of the etheric, and colours of the more subtle layers will emerge.

This may be due in part to the high concentration of rods in the periphery of the retina, which are highly sensitive to light, as opposed to the cones which are more sensitive to colour and concentrated in the fovea. This may explain why the coloured layers seem to appear after the white etheric layer. It is also possible to sense and affect these layers with ones hands, which gives an effect similar to the repulsion of opposing poles of weak magnets being pushed together.

While mainstream science has not documented humans consciously sensing subtle fields such as the aura, emerging research hints at surprising sensitivities. Human cells can respond to extremely weak electric fields, even when those fields fall below previously accepted detection thresholds.

Moreover, preliminary studies show that some individuals brains may unconsciously register changes in Earth's magnetic field. These findings suggest that the human sensory system may be more attuned to subtle electromagnetic phenomena than traditionally assumed.

The Beginning of Complex Life

Although life is thought to have begun within a few hundred million years of Earth's formation, definitive evidence confirms that stromatolites existed 3.7 billion years ago. However, the pivotal moment that allowed complex life to exist on Earth was a symbiotic relationship between a bacterium, likely an α-proteobacterium, and a host cell, likely a type of prokaryote called an archaeon, which absorbed it around two billion years ago.

These archaic ancestors of mitochondria were once free-living bacteria, until their destiny became forever entwined with our own. That they now sustain both our physical vitality, and our barely perceptible radiance is fitting, for they are both ancient and eternal flashes of primordial fire, shimmering in the depths of flesh.

Possibilities for External Symbiosis

Where mitochondria inhabit the interior of cells, plasma-based life may engage with humanity through the aura at some level. We have examined the possible mechanisms that may be employed for communication and energy manipulation, using resonance. These may be processes that are used to engage with complex life, when attached to the fields of resonant charge and light that extend beyond the skin.

One internal symbiosis created within our cells, the other external, created within our aura. One provides chemical energy and the creation of an external gateway to our mind, the other, once connected perhaps provides intuition, inspiration, guidance and influence. This possibility suggests a layered ecology of symbioses.

Inside us, the mitochondria turn oxygen and food into energy, creating a by-product of emanated electromagnetic radiation, a beacon for our radiation hungry plasma lifeforms, which may draw upon that same radiance, dependent on the type of energy we emit. In return, they may deliver subtle guidance as intuition, prophetic dreams and synchronicities.

Maybe they form lifelong bonds to our physical form, protective and sustaining. In a positive symbiosis, this might feel like heightened intuition, inspiration, or greater energy levels, and a wholly positive experience. A conscience for our independent actions; a positive guardian presence woven into the fabric of our existence.

Others may be parasitic, feeding on negative emotions and offering anxiety and depression in return, leading us down paths of destruction, amplifying base urges and emotions in ways that corrode rather than elevate. This description is phenomenological rather than diagnostic, reflecting reported experiences rather than clinical mechanisms.

In this not so positive relationship, just as radio signals can disrupt one another, a strong external field could create interference or noise in the aura, which may manifest as fatigue, confusion, anxiety, or depression. If the entity's field stabilises against the human aura through attachment, it may remain tethered, feeding on the host's bioelectric emissions.

This is a possible mechanism for both symbiosis and parasitism. We can sometimes sense this in others. There are times when one feels emotionally drained after even a relatively short interaction with certain individuals. They walk away feeling temporarily uplifted and leave one feeling sad, depressed or uneasy. As if they have consumed positive energy, leaving their negative energies and vibrations in exchange.

In effect, an energy vampire, preying on the refined, higher frequencies in a parasitic way, offering nothing negativity in return. This may be another process of energy harvesting for some negative plasma-based lifeforms. This hypothesis offers a new meaning for the old saying:

"As you sow, so shall you reap."

In other words, the lifeforms that attach through our energy fields, may resonate with the exact same vibrations and frequencies. If our energy frequencies are elevated and refined, they may not even perceive us. Just as the tuning fork effect will only vibrate forks tuned to the same frequencies.

Perhaps this duality is connected to the ancient motif of the angel and the demon on the shoulders, not simply allegory but a reflection of an unseen plasma ecology that mirrors the dualities within ourselves. The difficulty, of course, is that we do not yet know whether this relationship is protective, exploitative, or something that exceeds those categories.

The same beings who appear in some accounts as guardians or teachers may in other contexts manifest as indifferent observers or even as destabilising presences. Symbiosis here need not mean benevolence.

Lichens, gut flora, and viruses are all part of biological symbioses that include tension, conflict, and rebalancing. It may be that humanity's relationship with plasma intelligences is similarly complex, involving phases of guidance, indifference, and disruption over long timescales.

Fateful or Fatal Attraction

The parallel is remarkable. Mitochondria prove that life thrives on unlikely partnerships, they are a precedent for the idea that unseen companions may help shape our destiny. If bacteria can become the engines of our vitality, perhaps plasma beings are now entwined with our subtler dimensions. Where one symbiosis created the human body, the other may structure the human soul.

If plasma fields resonate, entrain, and interfere with one another, mutually amplifying, draining, or distorting, a plasma lifeform may couple to the aura, synchronising with the nervous system, altering mood, thought, or behaviour. Our brain-connected senses fire through flows of ionic charge, producing thoughts and emotions which are reported to be able to be sensed in the emitted faint light of the aura.

If these thoughts and emotions are effectively broadcast, surely they act as a beacon to an external plasma presence. Therefore, it is possible they could easily read and understand these delicate rhythms and signals and respond accordingly. This scenario suggests that humans are porous beings, open to communication in as yet largely untapped ways.

This may be an interface for infiltration and interrogation by subtle means such as phenomena historically interpreted as telepathy, remote viewing, or prophecy, leveraging quantum and non-linear time effects. Maybe we are not isolated as individuals but woven into a larger ecology of light. Just as bacteria inhabit our bodies and fungi lace the soil beneath our feet, plasma lifeforms may move through and around us, invisible yet entangled with our very existence.

Crucially, these possible symbiotic relationships do not unfold in an abstract spiritual vacuum. They appear to cluster in the same geophysical environments that favour plasma manifestation; charged storm systems, faulted quartz terrains, iron-rich cratons, and atmospheres already primed for ionisation. Whatever their motives, the watchers seem to meet us at the interfaces where our biology and their plasma-based existence are both most viable.

Shuffle off this Mortal Coil

At death, these interactions take on a further dimension. Hospice workers and family members have spoken of seeing a mist or haze rise from the body in the moments after life ends. Sometimes formless, sometimes carrying the faint outline of the person before fading away. In some accounts the haze momentarily resembles a double of the deceased, fading soon after.

Modern hospice research has acknowledged such reports as part of the wide spectrum of end-of-life experiences, alongside visions of light, feelings of peace, or the presence of unseen visitors. Scientific explanations point to dim lighting, expectation, or the moisture of breath in the room. Yet the consistency of such accounts, across centuries and cultures, demands at least some attention.

At the same time, clinical studies have recorded brief surges of gamma brain activity at the moment of death, a flare of synchronised rhythms in the seconds before and after the heart ceases. It may be no more than the brain's final storm as oxygen dwindles, but it may be the physical echo of the psyches release, the moment the subtle light-body slips free of its dense physical incarceration.

While currently there is no conclusive scientific proof of life continuing in some form after death, it is inevitable that each and every one of us will receive an answer to this question. Across mystical traditions, this inner light has often been treated as purely metaphorical or purely spiritual. Within the framework of the plasma entity hypothesis (PEH), a third possibility emerges.

The language of light may reflect early, pre-theoretical encounters with plasma intelligences whose presence was experienced both outwardly, as atmospheric or luminous forms, and inwardly, as shifts in perception, insight, or altered states. The metaphor of light is not arbitrary. It sits uncomfortably, but suggestively, close to the physics of radiance, ionisation, and electromagnetic structure.

The Seven Heavens

Traditions speak of what follows this final process of physical existence. Judaism, Christianity, and Islam all describe layered heavens, usually seven in number, through which the soul must ascend. In the Talmud, seven heavens are named, each with its own qualities and angelic hosts. The Qur'an tells of Muhammads night journey through seven levels, meeting prophets along the way.

The number seven recurs elsewhere in religious and

philosophical contexts, such as Hindu cosmology, Buddhist celestial planes, Gnostic ascent, and in Hermetic thought. It may be no coincidence that of the eleven proposed dimensions in M-theory, which unifies the five different string theories, only four dimensions unfolded to form our three dimensional universe.

The three spatial dimensions, length, breadth, height, and the fourth one which is linear time, that are perceptible to us. This would leave seven, remaining dimensions, curled up or *compactified* at the subatomic scale. If we look at the human aura as structured in seven layers, as many traditions have described. Then perhaps what we call death, is in fact a passage where those layers are linked to these compactified dimensions in some way.

Each one separating to be released back into its cosmic analogue. The ascent through seven heavens may be the journey of those seven layers returning to their source. To reinterpret angels in this context is not to diminish them but to restore their power. Accounts of radiant, winged beings may be true encounters with plasma intelligences of higher order, luminous and fearsome inhabitants of an obscured dominion.

They may have been seen at bedsides, guiding or receiving the departing soul, guardians of the thresholds between one realm and another. A guided ascent suggests that human beings are not simply extinguished at death but transformed by unburdening. Similar to an upturned glass underwater, the end of life would be like turning the glass upright, allowing the trapped air to be released.

CONCLUSION

We are bodies of flesh, intertwined with subtle energy. At the end, the subtle body rises, passing through layers of existence that the ancients called the seven heavens. Science has not proven this, but it has given us glimpses. Faint light from living tissue, surges of energy at death, reports of mist and colour.

Tradition provides the map, testimony provides the witness, and speculation bridges them into meaning. Perhaps in this confluence lies the truth. That to live is to be entwined in an invisible ecology, a hidden biosphere where plasma lifeforms reside, and to die is to fully rejoin them in those luminous realms beyond sight.

13 Controlling the Narrative

Since the first sightings of flying discs in the 1940s, through to the adoption of the term unidentified anomalous phenomena (UAP) today, official investigation has oscillated between curiosity and denial. The history of government involvement is not a straight line of inquiry, but a looping sequence of interest, suppression, and rebranding.

Behind this pattern may lie reasoning other than just military caution. Maybe the underlying reason is an attempt to preserve an ontological order, the worldview that underpins civilisation itself. In chapter one, I mentioned a fourth possible motivation for the UFO coverup. This chapter will address this line of thought and explore the reasons for it.

Alongside the familiar motives of technological secrecy, geopolitical advantage, and bureaucratic self-protection, I am suggesting a fourth and deeper motive for the management of UFO and UAP narratives: the preservation of ontological order. In other words, the defence not just of airspace, but of worldview. When we look back at the many times the US and other countries have made attempts to investigate the phenomenon of UFOs, the conclusions drawn always leave one feeling underwhelmed.

Thousands upon thousands of reports from all types of witnesses, and almost all with little to no apparent substance, or attributed to temperature inversions, mental illness, hallucination, or misidentification, and of course let's not forget the infamous swamp gas hypothesis. Was this all a part of the plan to make UFOs a taboo subject for general discussion?

I think there is very little doubt that is the case. Is the reason that this psychological control mechanism, for it most certainly was one, was adopted to hide extraterrestrial visitation and technology? Maybe not, maybe the truth would be a harder pill to swallow than extraterrestrial, little green men.

Project Sign and the Origins of Inquiry

In 1948, as the memory of the Second World War faded into the paranoia of the Cold War, the ~~United States~~United States Air Force initiated Project Sign, its first formal study of UFOs. The intelligence community was anxious to determine whether these objects were of Soviet origin.

Early reports were taken seriously, and Sign's analysts initially leaned toward the possibility of an extraterrestrial explanation. Their secret *Estimate of the Situation*, said to have supported this view, was suppressed and destroyed following sceptical review from the Air Force Chief of Staff. By 1949, Project Sign had been replaced by Project Grudge, which reinterpreted most reports as misidentifications, illusions, or hoaxes. An obvious attempt to control the UFO narrative.

Project Blue Book

In 1952, under public pressure and amid thousands of new sightings, Project Blue Book replaced Grudge. Over seventeen years, Blue Book collected more than 12,000 reports, including radar-visual events and pilot encounters.

Its *Special Report No. 14* (1955), conducted by the Battelle Memorial Institute, provided statistical analysis indicating that a significant minority of cases defied explanation and that the quality of the unknown cases was higher than the known, but despite these results, the report's summary downplayed their significance.

Blue Book became less a research project than a bureaucratic firewall between the phenomenon and the public. When it was finally terminated in 1969, the official statement was unequivocal:

"UFOs pose no significant threat to national security and offer no evidence of advanced technology."

Maybe this last statement was in fact correct if considered from the viewpoint of our plasma entity hypothesis (PEH). If this was the case, there would be no advanced technology, regardless of the entities' abilities. From the standpoint of the PEH, this historical reluctance to engage with the phenomenon is doubly ironic.

If a hidden plasma biosphere does exist, then the data gathered by these projects may already contain unrecognised evidence of it; structured luminous forms, altitude-dependent behaviour, electromagnetic disturbances, and repeated clustering over specific geological terrains.

Yet because these were filtered through an expectation of solid, engineered craft, anything that did not resemble a vehicle was pushed to the margins as either irrelevant, misidentified, or anecdotal. The very non-materiality that makes plasma life plausible also made it invisible to institutions looking for nuts, bolts, and propulsion systems.

The Robertson Panel

While Blue Book was in its infancy, the Robertson Panel, convened by the CIA in 1953, was tasked not with determining the reality of UFOs, but with managing their impact. Its scientists concluded that while the phenomena were not a direct threat, the public fascination surrounding them was.

They recommended a campaign of *education* and *debunking* to reduce what they termed mass hysteria. Civilian UFO organisations were to be monitored, and media outlets encouraged to treat the subject with ridicule rather than curiosity. From this point onward, disbelief became policy and the shift from scientific scepticism to social strategy was subtle but decisive.

The Condon Committee

In 1966, amid renewed public interest following major sightings over the ~~United States~~United States, the Air Force funded the University of Colorado UFO Project, led by physicist Edward U. Condon. Marketed as an impartial scientific study, it was intended to be the definitive word on UFOs. Internal correspondence later revealed that Condon had prejudged the outcome.

His final report (1968), often referred to as the Condon Report, declared that further study of UFOs:

"...probably cannot be justified in the expectation that science will be advanced thereby."

This conclusion effectively ended government-funded UFO research in the ~~United States~~United States. The scientific community, weary of ridicule, followed suit.

Yet buried within the report were cases that remained inexplicable, radar-visual tracks, ground traces, and electromagnetic interference effects that defied prosaic explanation. By ignoring the scientific anomalies, the Condon committee thus fulfilled its political purpose.

From Project Sign's tentative openness to the extraterrestrial hypothesis, through the studied scepticism of Project Grudge, to the bureaucratic caution of Project Blue Book, the pattern is consistent. Each programme acknowledged that something unusual was being observed. Each collected large volumes of data, some of it from highly trained witnesses.

Each, in its final accounting, steered public understanding away from ontological disruption and toward familiar reassurance: misperception, misidentification, atmospheric optics, psychological factors. What changed from report to report was not the phenomenon itself, but the official tolerance for ambiguity.

Parallel Programs

Across the Atlantic, similar investigations followed the same trajectory. The Flying Saucer Working Party (UK Ministry of Defence, 1951) dismissed reports as illusions and advised no further study. Two decades later, the UK's Project Condign (1997 – 2000) examined over ten thousand reports and concluded that some may be due to atmospheric plasma phenomena, an admission that luminous, energy-based anomalies do exist, albeit under natural conditions.

Condign's declassified report acknowledged that plasma formations could produce structured light effects, radar returns, and apparent manoeuvres, echoing features long noted in UFO reports. France adopted a more open approach. Beginning with GEPAN in 1977 and evolving into GEIPAN within CNES, the French space agency, it remains one of the few government-backed scientific programmes still active, publishing its case files and recognising a residual category of unexplained phenomena.

In Chile, CEFAA, part of the national aviation authority, has similarly studied pilot and radar encounters, publicly releasing declassified footage and reports. In the ~~United States~~United States, official attention returned quietly under new names. The Advanced Aerospace Threat Identification Program (AATIP) operated within the Department of Defence from 2007 to 2012, followed by the UAP Task Force, and later the All-domain Anomaly Resolution Office (AARO) established in 2022.

Each iteration claimed neutrality, yet their findings echo those of earlier eras; objects exhibiting unusual flight characteristics, unknown propulsion, and persistent radar signatures. The Office of the Director of National Intelligence (ODNI) released its Preliminary Assessment on UAPs in 2021, analysing 144 military encounters, 80 of which involved multiple sensors.

The report avoided conclusions but acknowledged that the majority remained unexplained. Two years later in September 2023, NASA released its own independent study, known as the UAP Independent Study Team Final Report, calling for open data and de-stigmatisation. After seventy - five years, the cycle repeats; renewed curiosity tempered by the same reluctance to confront the deeper implications.

If we consider these studies collectively, a pattern emerges. Each investigation begins in earnest, gathers compelling evidence, and ends with dismissal. Official policy oscillates between fascination and denial, suggesting not simple incompetence but strategic containment. During the mid-twentieth century, religion was inseparable from social order and post, Second World War governmental paranoia was rampant.

The notion that God, or at least his angelic emissaries might be plasma-based lifeforms; lifeforms so fundamentally different from our own physical form, was not a perception that would assist governmental control. The true, underlying motive may not have been to protect technology but to protect civilised society's belief systems, systems that underpin modern civilisation and prevent descent into social disorder.

Admission that such entities exist, and are linked back through history, even before Biblical times, would certainly dissolve the sacred divide between religion and physics, plunging society into disarray. Our human bodies, constructed of physical, or baryonic matter; theirs formed from ionised particles and crystalline lattices. In mid-twentieth century society, had full revelation of angels been publicised as tangible, existential lifeforms; and exposed as a multitude of species, mass panic may have ensued.

If that revelation had exposed them as insectoid, serpentine, or cetacean-like in form, this would surely have caused mass hysteria on an unprecedented scale and unhinged society, just as had been suggested by the Robertson Panel. It may well have undermined and upended theology, philosophy, and most importantly, governmental authority and control over the populace. As such, this was a colossal threat that had to be meticulously and skilfully handled.

A Lack of Physical Evidence

Through the whole of the extraterrestrial or UFO cover up, there has been no physical evidence that has been available for investigation; no craft, no alien beings and no technology that have been presented which would constitute proof of extraterrestrial visitation.

There have been many photographs and some very convincing testimonies from very credible witnesses. There is no doubt they are utterly convinced that what they saw and experienced is true and real, and I do not in any way wish to diminish the validity of their testimonies. The fact remains, they are second hand accounts, regardless of the character of the witness and the number of testimonies.

If this theory is correct, the instillment in the collective mind of society, of extraterrestrial visitations to our planet in technologically advanced vehicles, is masterful. Fuelled by disinformation and many sightings of moving lights in the sky the extraterrestrial hypothesis soon found a home in the public consciousness. Denial, ridicule and threat perpetuated the myth, growing it exponentially as the years passed.

The human mind is able to be manipulated in many ways, we know this. Fear and intimidation, hypnotism, narcotics, advertising, and even certain frequencies can affect our perspective of reality, and we also know there have been programs that have looked to exploit this.

Disinformation agents are known to have been employed, and we may never know to what extent they may have altered the public perception. There is also the use of threat, bribery and blackmail as tools of control. We must also not discount the possibility that secret government programs may have advanced through our own, human efforts, maybe yielding exotic technologies, leading to craft sightings, further muddying the water.

Many reports of UFOs and UAPs, including those over nuclear missile silos and military bases, describe luminous discs, fireballs, orbs, tic-tacs or cigar shaped craft. Descriptions that are endemic to plasma entity phenomena, but which are almost always cited as physical *craft*. If these were physical craft, what better way to expand the deception than to have sightings by qualified expert military personnel.

The Kenneth Arnold Sightings

On June 24th, 1947, Kenneth Arnold reported sighting nine, shiny UFOs. near Mount Rainier, travelling in formation at an estimated one thousand nine hundred and thirty two kilometres per hour, or one thousand two hundred miles per hour. This is often referenced as the first, post-Second World War UFO sighting, and the beginning of the modern era of UFOs. However, later in life, Arnold changed his theory, after what he believed were seven further sightings, instead portraying UFOs as being:

"...groups or masses of living organisms that are as much a part of our atmosphere and space, as the life we find in the oceans."

Conclusion

The Condon committee and its predecessors appear less as failures of science and more as acts of narrative preservation and upholding methods of control. They maintained the theosophical view and concluded that humanity is alone, a view necessary for a stable worldview and controlled society. Was the creation of the extraterrestrial hypothesis part of this elaborate strategy, designed to conceal the existence of a potentially destabilising revelation?

The Roman military strategy of "*divide et impera*" or "divide and conquer" comes to mind. Create a fantastical false narrative, stigmatise that narrative to create division and perpetuate the storyline with disinformation and mind control. It may seem far-fetched, especially if one is invested in the extraterrestrial hypothesis (ETH).

Yet as plasma and consciousness research, and high-resolution imaging continue to advance, the illusion thins. If the proposed hidden biosphere exists, then history's committees and cover-ups were not suppressing fantasy, but managing revelation, delaying the moment when the world must face that the divine is luminous, intelligent and not what we were told.

The Condon committee and its counterparts may have managed to close the door on open inquiry into the UFO problem. However, this was not a new problem, it was inherited. The impulse to define, delimit, and domesticate anomalous phenomena has accompanied civilisation from its earliest beginnings. Every age has its committees and inner-circles, keepers of esoteric knowledge and information.

The ancient world called them priesthoods; the modern world calls them scientific panels, but both serve the same function; to determine which experiences are permitted to exist in public reality. The same instinct that silenced anomalous light in the modern age once sanctified it in antiquity. Where committees now redact, priesthoods once canonised. The following chapter traces that earlier order of containment, the theological bureaucracy that first taught humanity to mistrust its own eyes.

For a civilisation whose dominant religions have long rested on revelations mediated by radiant messengers, wheels of fire, pillars of cloud, and winged beings of light, the prospect that such figures might be expressions of a non-biological plasma intelligence is profoundly destabilising.

It is one thing to admit that unidentified craft violate airspace. It is quite another to admit that the very category of "spirit" or " angel " may have been an attempted description of a neighbouring lifeform in an electromagnetic ecology we barely understand. To acknowledge the latter would not only challenge national security paradigms, but unsettle the metaphysical scaffolding of entire societies.

14 Origins of Theocracy and Coverup

This chapter does not claim definitive proof of the physical nature of the entities described in ancient texts, nor does it assert that religious authors possessed a scientific understanding of plasma physics.

Rather, it adopts a comparative phenomenological approach, examining recurring descriptive motifs across disparate historical periods, cultures, and interpretive frameworks. The focus is not on theology as belief, but on testimony as record.

By comparing ancient narrative language with modern observations of luminous, structured, and apparently autonomous atmospheric phenomena, this chapter explores whether consistent perceptual features may underlie radically different cultural explanations.

It further examines how institutional systems, religious or governmental, tend to stabilise disruptive phenomena through translation, classification, and semantic reframing. In this sense, the chapter is concerned less with what these phenomena ultimately are, and more with how they are rendered intelligible, and controllable, through language.

Long before the first radar return or committee memorandum, there were descriptions of luminous intelligences moving within the air, gods, messengers, hosts of heaven, and living fire. These were not vague symbols. They were detailed reports of structured, radiant forms: rotating, layered, and often accompanied by wind or vibration.

Yet from an early period, these encounters were absorbed into controlled narratives. That which could not be denied was sanctified; that which could not be sanctified was silenced. Subtle variations in translation can be found in different versions of biblical texts, which were then built upon to contain this information.

The editorial response to the unknown was to name and classify. Angels multiplied where witnesses once spoke simply of aerial lights or living fire. Their domains were divided, their functions enumerated. The result was a celestial bureaucracy; a heavenly mirror of the institutions that required obedience on Earth.

By the fifth century, when Pseudo-Dionysius systematised the nine choirs of angels, the heavens had become an ordered empire. Seraphim, Cherubim, Thrones, Dominions, Virtues, Powers, Principalities, Archangels, Angels; each rank assigned a purpose, each purpose anchored to doctrine.

No new order could be admitted without ecclesiastical consent. The process was identical in motive to that of the twentieth century's committees; containment by redefinition. Theologians transmuted radiant beings into metaphors of grace, just as government officials translated structured light into radar error and psychological projection. In both cases, semantics of language functioned as the veil.

THE BOOK OF ENOCH

To recover what may lie behind that veil, one must return to the texts before systematisation. A time before the use of these redefining metaphors. The Book of Enoch provides us with such a window.

Composed between the third and first centuries BCE, the Book of Enoch predates Pseudo-Dionysius by half a millennium, while Ezekiel's vision lies still earlier in the sixth century BCE. Between them runs an unbroken thread of radiant visitation.

The following passage is from a commonly used translation of Enoch's description (R. H. Charles, 1912) of Chapter 14, verse 8:

"...and the vision was shown to me thus: Behold, in the vision clouds invited me and a mist summoned me, and the course of the stars and the lightnings sped and hastened me, and the winds in the vision caused me to fly and lifted me upward, and bore me into heaven."

The use of the term "*vision*" (Hebrew-ḥizayon and Aramaic-ḥazon) in apocalyptic literature often designates a revelatory experience, which may include sensory-rich descriptions. The language here is vivid and physical; "*clouds invited me*" and winds, "*...caused me to fly and bore me into heaven.*"

These suggest more than a dream or mere metaphorical vision of the mind, the text frames it as if Enoch experienced an encounter of invitation, transport and arrival. Ancient audiences may have made less of a distinction between a vision of travel and actual travel, as truth, myth, and sensory experience often overlapped and blended. The term "*vision*" in translation does not necessarily imply a purely symbolic reading and leaves open the option that the writer describes an experience he regarded as real.

Looking at different renderings shows nuanced variations in translation:

"...clouds and a mist invited me; agitated stars and flashes of lightning impelled me; winds in the vision assisted my flight."

Note that the translator preserved "*vision*", which softens the reading, whilst the description itself remains vivid and transportive, and the difference between "*invited*" versus "*summoned*", "*lifted me upward*" versus "*bore me into heaven*". The verb choices and prepositions change how literal or metaphorical the ascent seems.

Some scholars suggest we could read the Hebrew/Aramaic more literally as "clouds called me" and "*a mist separated me*", which emphasises an external agency or forces in the sky rather than an internal psychological experience. Essentially, the text can be read as, "*I was brought*" rather than, "*I imagined*".

Translation is itself a containment mechanism. Each linguistic tradition, Hebrew, Greek, Latin, English, imposes its own metaphors of obedience and purity. What was once a cloud of fire, becomes glory; what was movement, becomes a vision; each substitution deflects the literal toward the devotional. Enoch's description continues:

"...And I entered in until I drew nigh to a wall which is built of crystals,

and surrounded by tongues of fire and it began to affright me. And I went into the tongues of fire and drew nigh to a large house which was built of crystals and the walls of the house were like a tessellated floor made of crystals, and its groundwork was of crystal."

If we examine this and the previous passage from a context of the plasma entity hypothesis (PEH), we have an encounter where:

- Enoch is invited, possibly telepathically, by intelligent clouds: Manipulation of cloud and apparent telepathic interaction with humans are abilities associated with plasma beings.

- Then "*agitated stars*". This may suggest that what seem to be stars, are flickering more than stars normally flicker due to the effect of the Earth's atmosphere. This has been repeatedly observed and associated with plasma lifeforms.

- "*...and flashes of lightning impelled me*". Lightning, as we know is a natural form of plasma. Plasma entities are theorised to "feed" on this energy and have often been observed in and around lightning storms.

- Enoch is then, "*lifted by a wind*". Some translations specifically state a whirlwind. This could be a magneto plasma conduit, a channel linking the ground to a higher plasma region in the ionosphere. Plasma entities are observed and associated with wind vortices as earlier discussed in the chapter on willy-willies (dust devils).

- Enoch then describes how the winds, "*...bore me into heaven.*" This may have been Enoch's perception of reaching the boundary of an ionospheric or high-altitude luminous zone, or above cloud level on a physical mountain. Mount Ararat, a stratovolcano, at over 5,000 metres (16,850 feet) is a possibility, although Enoch's location is not stated.

- Enoch then entered in and came to, "*...a wall built of crystals...*" Once again, we have a description of geometric crystalline forms, associated with plasma-based lifeforms and complex dusty plasma.

- "*...surrounded by tongues of fire.*" The tongues of fire frightened Enoch, but he entered into them without harm which suggests structured plasma or coherent light-forms. The word fire would be the only terminology available to Enoch for what he saw.

- "*...to a large house which was built of crystals: and the walls of the house were like a tessellated floor made of crystals, and its groundwork was of crystal.*" A large crystal structure with crystalline walls of neatly fitting geometric shapes in repeating patterns, although some translations state stones of crystal or fitted crystal stones, with a foundation also formed from crystal.

Is this a UAP encounter from around five thousand to six thousand years ago, using language that can be read like a physical experience? Through translation it's meaning is slanted towards this encounter as being a vision.

Sceptics may argue that such passages reflect inner vision rather than outer event, yet the precision of physical description, directional wind, crystalline architecture, structured light, suggests an empirical core beneath the devotional surface.

Could this skewing of the described events be designed to

obscure the true nature of the divine? It is certainly a possibility. Every part of Enoch's encounter corresponds to known plasma behaviours and plasma entity observations. Although the author calls it a vision, the language betrays a belief in a physical experience. This aligns with what one would expect from an encounter with high-energy plasma phenomena rather than a mere metaphor.

Enoch's encounter is not an isolated incident. Ezekiel's vision repeats the pattern with greater precision: a "*whirlwind out of the north*", a "*great cloud*", and a "*fire infolding itself*". Four radiant beings appear like "*burning coals of fire*", an apt description of many plasma entity sightings, their movement accompanied by "*wheels within wheels*" and their "*rings were full of eyes round about them four*", this is a telling passage.

The concentric rings of a plasma orb, infolding and with what are described as, "*full of eyes round about them four*", corresponds perfectly to images captured which show black spots which could indeed be mistaken for eyes. These recurring motifs reveal not artistic coincidence but perceptual continuity; the same visual grammar applied to differing ages of technology.

The Hebrew allows several readings of this description, as the Ophanim are described as wheel-like beings, but may also be rings, rims, or toroidal in shape. The fire infolding itself describes precisely the inward-turning eddies of magnetised plasma, often seen in plasma entity orbs. In modern terms, the description fits a rotating plasma structure, whose surface roils and churns and is surrounded by moving magnetic poles.

This vision then, is not purely mystical but phenomenological; an ancient account of plasma behaviour translated into the vocabulary of sentience. Later interpreters, uncomfortable with its physicality, converted it into the mystical Merkabah, a chariot of meditation rather than a chariot of fire.

The Appearance of Fire

Witnesses in every age have reported luminous spheres. Their colours, dull red and amber through to white, mirror the thermal and spectral range of low-temperature plasma. Observers have noted internal rotation, surface flicker, and side-darkening, features equally familiar in atmospheric plasma and also optical diffraction.

Whatever their origin, the visual constants are ancient; a glowing sphere that seems alive, its centre roiling, its light folding back upon itself. "Burning coals of fire", Ezekiel called them. This term is as apt for modern sightings as it is for Bronze-Age vision. These ancient descriptions are what they saw, and later, scribes supplied theology.

As sceptics will remind us, the human eye itself can add to the mystery. Corneal reflections and diffraction rings create entoptic halos around bright lights, which can also produce the illusion of layered wheels. Yet even if physiological optics contribute, the perception remains accurate in form. The "wheels within wheels" description, is the natural geometry of light passing through curved media.

Language as Containment

The codification of angelic types completed the containment. What began as unsorted observation became an ordered hierarchy:

- Seraphim, the "burning ones", may have been a description of the visible orb.

- Cherubim, multi-faced winged constructs, may be a record of the geometric cuboid structure within the plasma orb, perhaps seen from differing vantage points.

- Ophanim, the "wheels full of eyes", preserve the memory of polarised, toroidal plasma.

By naming and ranking them, religion absorbed them into its control system. No longer autonomous intelligences, they became servants of dogma. This theological domestication set a precedent. When twentieth-century institutions faced renewed evidence of luminous intelligences, objects that behaved with apparent volition, yet defied material classification, they too turned to linguistic defence.

Terms like unidentified aerial phenomenon, temperature inversion layer, and psychological misperception replaced angel, seraph, and watcher. The underlying event, contact with structured luminosity remained constant, only the interpretive framework changed.

Ancient and Modern Skies

Modern observers note that plasma orbs often display concentric rings, turbulent centres, bordered by spots or patches of darkness. These features are explicable within plasma physics; sheath layering, magnetic dipoles, and self-shadowing, yet these same features also match the ancient lexicon of eyes, faces, and wheels or rings.

Whether the entities are ultraterrestrial, extraterrestrial, or interdimensional, the literary act of anthropomorphisation incorporates luminous configuration into spiritual form and becomes an instrument of control. It is a continuous thread from the "*watchers*" of Enoch to the debriefing rooms of secret government departments. The question is not whether governments or religions lie; it is how they translate. The translation itself enables the mechanism of authority.

The Politics of Heaven

To control the narrative of the sky, is to control the narrative of existence; the ancients knew this. Whoever defined the heavenly order also defined the moral order below. The same holds today; to define what is real in the upper atmosphere is to delimit the boundaries of the possible.

Hence the convergence of theology and bureaucracy, both seeking to preserve stability by owning interpretation. The theologian's canon and governmental reports are expressions of the same instinct; to protect the public from the disruptive implications of direct experience. One canonises, the other classifies, both reducing the incomprehensible to administration.

The Hidden Continuum

If we strip away those successive interpretations, what remains is a continuum of phenomena consistently and reliably described across millennia:

- *Whirlwind*: An electromagnetic vortex or plasma conduit.
- *Wheels within wheels*: Concentric toroidal rings.
- *Eyes round about*: Luminous nodes or magnetic dipoles.
- *Burning coals*: Red or amber plasma temperatures.
- *Bright cloud*: Ionised envelope or corona.

Comparable cross cultural themes of fiery chariots, descending wheels, and luminous beings, reappear in Mesopotamian, Vedic, and Mesoamerican accounts, implying shared observation. Whilst these translations do not prove equivalence, they do demonstrate consistency. Humanity has been witness to a coherent class of phenomena since records began, the difference lies only in semantics.

Toward Disclosure of Meaning

Perhaps the deepest concealment is not institutional but linguistic. Angel, UAP, entity, weather phenomenon, each is a title of containment and to name is to domesticate. True disclosure, then, would not arrive as a press conference but a restoration of perception; the capacity to see this hidden realm without inherited filters.

What has long been institutionally concealed, either deliberately or through habit, is that celestial inhabitants are still visible. Not as an abstract, but as structured, intelligent, life, moving through the atmosphere as living plasma, alive in the true sense of the word. To acknowledge that would blur the boundary between heaven and earth, between science and spirit and would end the monopoly of interpretation.

Conclusion

This hidden theocracy is therefore not a secret religion but a system of linguistic intervention. Across epochs, the administrators of reality, priests, theologians, scientists, and intelligence officers, have inherited the same task: to mediate between humanity and the light that surrounds it.

Each generation re-writes the lexicon. The phenomena become angels, the angels become data, and the data are filed away. Yet in the space between the classifications, the light still moves, wheels within wheels, eyes within fire, awaiting recognition as both ancient and alive.

PART TWO

Exploring the Phenomena

15 Entities and Aircraft

On the 2nd of January 2024, just a couple of months into my investigation, while my partner and I were visiting friends in the Perth Hills, I noticed a commercial aircraft in the distance and grabbed my mobile phone. I filmed the aircraft as it approached us and passed overhead, en route to Perth Airport.

I finished filming and slipped the phone back into my pocket. It was only later, when reviewing the video, that I noticed two large luminous orbs just above the starboard wing, as if they were hitching a ride. The sight, although incredible, initially unsettled me. I felt intense curiosity tinged with unease, as I fly regularly to work in the Pilbara.

I repeatedly played the video, expecting the orbs to somehow change into something recognisable as I had formed no hypothesis at that point. I entertained the possibility that the orbs were artifacts caused by refraction or reflection, but I don't believe that to be the case, in light of subsequent observations.

A few weeks later, at a winery in the Swan Valley, I filmed more aircraft on their approach to Perth airport and discovered other anomalies. In one video, an object almost matching the size of the engine, seemed to cling to the engine casing below the starboard wing. I could not help wondering whether these beings accompanied every aircraft, unnoticed by the naked eye.

I decided to dig deeper into this new expression of weirdness, and one cold evening in mid 2024, armed with the Sionyx camera, I spent five hours in the car park of that same winery, waiting for aircraft flying to and from Perth airport. The capture I was most excited about from that outing was of a northbound commercial passenger aircraft just after take-off, which passed overhead.

As it did so, a white, translucent, amorphous shape appeared near the nose of the aircraft. It then swiftly stretched out like a glowing slinky toy, as it rapidly moved the length of the fuselage, snapping back to its original amorphous size and shape at the tail end. It then disappeared above the tail of the aircraft.

As it did so, a second amorphous orb emerged from above the port wing, following the first along the underside of the fuselage. These were not reflections, nor were they artifacts within the recording. They were real objects, observed only through the lens of the full-spectrum camera, moving independently and seemingly with purpose along the underside of the aircraft.

I took some screenshots, adjusted the images to reveal as much detail as possible of both objects. The first object gave a clearer view being slightly larger and had a slight hint of pulsing colour, similar to the many orbs previously observed on video from the same camera.

The screenshots showed a slightly tapered anterior. Whether this was the "head" is unknown, as head may not be applicable. After some time researching and contemplating what these entities could be, the word that came to mind was, Gremlins.

GREMLINS

In recent times, the word gremlin conjures images of the 1980s children's movie of the same name, which tends to trivialise the subject. However, the gremlin is not a new concept and originates from much older sources. The word gremlin may derive from the Old English word, "*graimen*", which means to vex, or possibly from the more familiar, *goblin*. Whatever its origin, it gave name to the nameless.

Once mankind left the ground under powered flight, the term became part of aircrew humour and lore, passed around in mess halls and briefing rooms, appearing even in cartoons and training manuals. The myth long predates powered flight, but it was in the 1920s that RAF pilots stationed in Malta, the Middle East, and India began applying it to the strange inexplicable misfortunes of their aircraft systems. Instruments went awry, engines faltered, and compasses spun.

The cause was described as small creatures clinging to wings, tugging at control wires, or grinning as they jammed machine-guns. This folklore may have offered psychological comfort, a way of anthropomorphising misfortune, yet the persistence of reports, and the resemblance to my own recordings of objects attached to engines and wings, suggested that something more than metaphor was at work.

FOO-FIGHTERS

There is no era in modern history more synonymous with

strange aerial phenomena than that of the Second World War. Many sightings from aircrews, both allied and axis, were documented. In November 1944, over the Rhine, pilots of the 415th Night Fighter Squadron reported a formation of eight to ten fiery orbs.

The orbs were orange, glowing and paced their Bristol Beaufighter. They moved with intelligence, mirroring the aircraft's manoeuvres, before vanishing as suddenly as they had appeared. The squadron dubbed them "*foo fighters*", borrowing from the surreal humour of the American comic strip Smokey Stover, where the nonsense word *foo* stood in for fire.

Within weeks, Associated Press correspondent Robert C. Wilson carried the story to the public, and Time magazine followed with a feature on 15 January 1945. Foo fighters became wartime news. The phenomenon was widespread as B-29 crews over the Pacific described "balls of fire" shadowing their bombers. RAF pilots over Germany reported "red spheres" weaving between their formations.

Even German and Japanese airmen told of strange lights pursuing their aircraft. In every theatre of the war, the orbs appeared, untraceable by radar, immune to gunfire, and seemingly inquisitive but uninterested in causing harm. The immediate suspicion was enemy weaponry as rumours swirled of German Feuerball devices, radio-controlled incendiaries, or plasma-based flak mines.

Yet no evidence of such weapons surfaced, and still the orbs showed no hostility. Naturalistic explanations were also attempted. St. Elmo's Fire, an electrical corona discharge, was familiar to sailors and pilots, but its bluish glow and attachment to sharp edges did not match the fiery spheres modus operandi.

Ball lightning was invoked as a cause, though ball lightning remains rare, transient, and poorly understood. Others suggested combat stress and hallucination, yet foo fighters were witnessed by entire crews, often across multiple sorties. Gremlins were often dismissed as folklore and foo fighters as mass hysteria, but these explanations fell short then, and they still fall short today.

In an environment where technology was fragile and danger imminent, it may be that it was easier to blame an invisible sprite than to shoulder all responsibility. However, I now had another theory; one that gave credence to the reports. From my own observations, it appears that aircraft act as attractive beacons for atmospheric entities.

A modern jet is not simply a mechanical body of aluminium and steel but a veritable smorgasbord of radiant energy across many frequencies. Ignition systems generate broadband radiofrequency noise, generators produce low-frequency harmonics, and the digital circuitry of control units throws out bursts in the gigahertz range.

Exhaust gases at temperatures between six and fourteen hundred degrees Celsius, shine brightly in the infrared spectrum, while the roar of turbines produces infrasound that can be sensed over long distances. Alongside these are the constant chatter of communications, radar pulses, satellite navigation signals, and even the hum of passenger Wi-Fi. To us, these emissions are invisible background radiations of frequency.

To a plasma lifeform or an entity attuned to electromagnetic gradients, they may be a signal for a feast of stimuli. Civilian airliners broadcast abundantly but military jets less so, for their designers strive to cloak them with shielding and emission controls. Yet suppression is not the same as elimination and even stealth aircraft leak their signatures.

Early Radar Frailties

The foo fighters of 1944 and 1945 were almost never confirmed by the myopic radar of the era, as the technology was simply too crude. Early sets were fixed to narrow bands, vulnerable to clutter from weather and terrain. Small or transient anomalies were dismissed as angels or atmospheric ghosts.

Angels, phantoms and *ghosts* were the strangely apt common terms used by operators for insignificant, faint, or transient returns on their radar screens that they could not correlate with known aircraft. From the standpoint of our plasma entity model, the phraseology seems poetically prophetic.

Plasma adds another complication, as a dense plasma cloud may reflect radar like a mirror, but a sparser one may allow the waves to slip through. Only when the frequency resonates with the plasma's natural oscillation does a strong return appear. Foo fighters may have existed in this liminal state, occasionally visible to the naked eye but mostly invisible to the radar systems of the time.

Modern Radar Complexities

Modern systems have transformed that picture. Signal processing now preserves what once was discarded as noise, pulling faint or erratic tracks from the clutter in the ether. Electronically scanned arrays send out beams across multiple frequencies at once, shifting them in microseconds, so that even a plasma orb tuned to one narrow band is likely to betray its existence.

Photonic radar, still in its infancy can, in principle, resolve objects only centimetres across, small enough to distinguish a luminous globe moving at altitude. Passive radar, which watches the way television, radio, or cellular signals scatter, does not need to emit anything at all, but can sense even the faintest ripple in the electromagnetic field.

Combined with infrared cameras, ultraviolet detectors, and optical sensors, these advances mean that phenomena once glimpsed only by pilots' eyes can now be logged, tracked, and confirmed. It is no coincidence that recent UAP encounters, such as those of the USS Nimitz or the USS Omaha, involved small luminous objects tracked both visually and on radar.

The foo fighters of the Second World War may not have disappeared at the end of the war. Instead, our technologically advanced instruments have finally begun to see them. Why did pilots describe gremlins as impish saboteurs and foo fighters as fireballs? A viable explanation may be because human beings interpret the unknown through the symbols at hand, just as in the biblical descriptions of Moses, Enoch and Ezekiel.

For the RAF, steeped in British folklore, it was natural to imagine mischievous sprites tampering with engines. For American airmen, immersed in comic strips and wartime slang, glowing foo fighters made sense. The act of naming the unnamed, offered a measure of psychological safety. To share a laugh in the mess hall about gremlins, or to swap tales of foo fighters at briefings, was to make the unknown less alien and more human.

Yet behind the humour, the uncanny truth persisted; something was accompanying them in the skies. What if these reports were real accounts of the type of plasma-based entities, I had filmed on the full-spectrum camera? What if their energy requirements attracted them to aircraft?

In modern reports of nearby encounters with UAPs, electrical interference of technology is often described. Engines of automobiles become unable to start, mobile phones and cameras producing strange effects, becoming hot and batteries rapidly depleting. The phenomena described by RAF pilots in the 1920s, by Allied and Axis crews in the 1940s, and captured on my cameras in 2024 all share common features.

They appear upon and alongside aircraft, interact with instruments and engines, and move with an intelligence that feels reactive rather than random. Had the radars of the 1940s been more advanced, foo fighters might be catalogued today as an early wave of UAP. Had full-spectrum footage been available to those pilots, gremlins might not have been dismissed as folklore.

Viewed together, they reveal a continuity across a century; humanity has never flown alone. Intelligent plasma lifeforms, drawn to our machines, have paced us from the dawn of powered flight to the present day. We must ask why they are attracted to aircraft?

The Allure of Aircraft

Many frequency emissions from multiple onboard systems are broadcast from commercial aircraft, as are electromagnetic emissions from engine systems:

- Engine-related frequencies can be registered from Ignition systems where pulsed, high-voltage discharges generate broadband RF noise typically in the kHz to low MHz range.

- Generator systems emit harmonics in the 50 Hz – 400 Hz (AC frequency) range and can produce switching noise up to several MHz, especially if power electronics like inverters are used.

- Engine Control Units (ECU) are systems on aircraft that are used to control various engine functions. They emit digital switching noise in the MHz to GHz range depending on the processor and data bus speeds.

- Exhaust temperatures of between approximately 600° C to 1,400°C, emit thermal or infrared (IR) radiation peaking between 3 μm – 5 μm and extending into the 8 μm – 14 μm range, which are detectable by infrared sensors.

- Acoustic waves in the infrasound range are less than 20 Hz and typically between 5 Hz – 16 Hz, which are detectable tens of kilometres away.

- Additional engine noise can be from blade-passing frequencies which are often in the 500 Hz to several kHz range and combustion noise, which is detectable as low-frequency broadband noise, typically centred around 100 Hz – 500 Hz.

- Communication systems on commercial aircraft for Air Traffic Control (ATC) voice communications use VHF 118 Hz – 137 MHz.

- Weather radar uses the X-band which is 8 Hz – 12 GHz.

- DME systems measure distance for navigation using approximately 960 Hz – 1215 MHz.

- Transponders (ADS-B) are used for aircraft tracking and use an interrogation frequency of 1030 MHz and reply on a frequency of 1090 MHz.

- GPS Receivers for Positioning using 1.57542 GHz (L1) and 1.2276 GHz (L2).

- If Wi-Fi is available for passenger connectivity, it will broadcast on 2.4 GHz and 5 GHz.

- Inverter and power systems produce electrical noise in the frequency range of 20 kHz – 1 MHz.

As can be seen from this quite substantial list, there are many frequencies of emissions from commercial aircraft, any or all of which may be attractive to plasma or energy based lifeforms. Let us also look at how military aircraft actively reduce and suppress frequency emissions, by using shielding and emission reduction techniques to reduce the electromagnetic signature.

These technologies are designed to avoid detection by enemy forces surveillance technologies. How effective these techniques are in shielding the aircraft from energy based lifeforms is unknown. The various techniques used by military aircraft, that are public knowledge, are intended to block or absorb signals which include:

- Faraday cages and shielded wiring and connectors which are installed around avionics and reduce RF leakage and interference.

- Modern radars like AESA have Low Probability of Intercept (LPI) systems which use spread-spectrum signals, frequency hopping and stealth waveforms which are designed to be hard to detect or jam, even when active.

- Electromagnetic (EM) suppression units to reduce ignition and generator emissions and infrared signature suppressors can also mask the heat signature of the exhaust.

- Intentional Emission Management (EMCON) protocols allow aircraft to fly in near radio silence, disabling or minimising systems such as radios.

- IFF (Identification Friend or Foe), Data links, and Radar. These EMCON protocols are used to avoid detection by enemy radar, electronic surveillance and infrared tracking systems.

- Signature Reduction Strategies make the aircraft harder to detect by enemy surveillance technologies.

- Radar-absorbent materials (RAM) reduce reflection.

- Aircraft shapes are designed to minimize the radar cross-section (RCS).

- Infrared (IR) masking of exhaust vents and coolant systems using mixed air, lowers plume temperature.

These complex emission signatures make modern jet aircraft hypothetically detectable and attractive to entities sensitive to electromagnetic (EM) gradients, infrasound, or thermal contrast and whilst military jets shield, suppress, and strategically control their frequency emissions to avoid detection, jamming and targeting, they cannot be eliminated completely.

Plasma Lifeform Diversity

Foo fighters and gremlins are possibly just two of many species of Earth's hidden biosphere, or possibly two separate manifestations of the same type of entity, with differing morphologies, specifically required to satisfy diverse energy requirements.

The variety of manifestation of the phenomena suggests they may also be known by other names, in other cultures and other time periods. The diversity of form in this hidden realm may well parallel that of biological life on Earth. According to a 2011 study, 86 % of land species and 91 % of marine species are still unknown.

More recently, large-scale ecological modelling has suggested that Earth may host in the order of $10^{11} - 10^{12}$ microbial species, implying that the overwhelming majority remain undocumented. At present, our knowledge of these lifeforms is limited, and we can only speculate on their origins, their requirements and their intentions.

Their apparent attraction to diverse electromagnetic frequencies and radiated energy is apparent as we will explore in other chapters. Gremlins and foo fighters were not simply inventions of tired minds or high-altitude psychosis. They were names given by aviators, to encounters with a plasma reality that lay outside the framework of their time.

Today, as radar, optics, and infrared begin to confirm what

eyes alone once witnessed, we are left with a sobering realisation that our atmosphere is not empty. Alongside our aircraft exist entities of intelligence, light and energy, unseen by most yet undeniably present. The Second World War air campaign revealed them unintentionally and they remain here still, inhabiting the sky.

16 Food of the Gods

The New Jersey Drone Incursions

In chapter six we explored the possibility that the South Atlantic Anomaly (SAA) may have been a mechanism by which plasma-based lifeforms, if they are interdimensional, may have entered our dimension after Earth's magnetic field had been weakened by the May 2024 solar storm.

It was hypothesised that these lifeforms could then have travelled north, attracted by the nearest beacons of electromagnetic and nuclear radiation; the Eastern seaboard of the ~~United States~~United States.

In this chapter we look at that scenario once they arrived. In late 2024, waves of what were described in mainstream media and US government statements as drone incursions appeared across the ~~United States~~United States. More specifically along the east coast, over densely populated areas and sensitive military installations.

Since November 2024, reports of sightings have caused alarm and consternation in the ~~United States~~United States general public and government alike. During the initial series of sightings, over five thousand reports were made to the FBI, which prompted multiple agency investigations and widespread media coverage.

On the 12th of December 2024, erroneous reports of a crashed drone, were in fact of a Tecnam P-2008 aircraft, which crashed on I-684 in Greenwich, Connecticut due to engine failure. At the time this seemed like an attempt to strengthen the drone narrative and confirm that the objects in the skies over the USA were just that, drones.

Official Incursion Responses

Some of the theories put forward in an attempt to explain these sightings appeared to be founded on paranoia or wild speculation. As the sightings continued unabated, with no definitive answers forthcoming, or any means of halting the nightly intrusions, the public perception was that the US military could not defend its airspace.

Among the more serious proposals was the notion that the incursions represented hostile surveillance operations by foreign adversaries, using drones launched either from China or Russia. These were said to be probing US airspace with advanced unmanned systems. Another theory suggested that the drones may have been deployed from an Iranian mothership stationed offshore, launching sorties inland from international waters.

More mundane explanations were also offered, including the possibility that these were FAA-approved commercial drone flights, operating under legitimate licenses but misinterpreted due to their unusual flight patterns. Likewise, some officials considered whether hobbyist drone operators might be responsible, perhaps unaware of the sensitive nature of the airspace they were encroaching upon.

Finally, a common fallback explanation surfaced; that many of the sightings were simply misidentifications of manned aircraft mistaken for drones by members of the public, unfamiliar with aerial silhouettes or nighttime operations.

According to reports, the drones, which some witnesses described them as being "the size of an SUV", were often observed hovering or manoeuvring at altitudes around 1,000 metres and for durations that stretched into several hours.

Such sustained aerial presence immediately rules out hobbyist, commercial, and even most law enforcement drones, which simply do not possess the energy capacity to remain airborne for that long at such elevations.

Technologies for Extended Flight Times

Drones would require a power source far more advanced than the lithium-ion batteries found in consumer models to be able to achieve the kind of endurance and altitude displayed by these objects.

One candidate to achieve extended flight times might be lithium-sulphur battery technology. Although officially experimental and still confined to the realm of prototype research, this technology offers a theoretical energy density of around 500 Wh/kg, but has yet to see practical application in long-endurance drone systems.

Another possibility is hydrogen fuel cell technology. This has already undergone testing in both military and certain commercial applications and boasts an impressive energy density ranging from 1,000 to 3,000 Wh/kg. Its added advantage lies in producing a relatively low thermal signature.

However, the use of hydrogen introduces its own complications. Safe storage, whether as compressed gas or within chemical hydrides, presents logistical and safety challenges. In addition, fuelling infrastructure for such systems is not widely available, particularly in clandestine or mobile deployments.

Another option is solar power, though while this technology is appealing in theory, it is naturally constrained by the availability of sunlight, rendering it impractical for nocturnal operations, which was when many of these sightings occurred.

A more grounded alternative is a hybrid gasoline and battery system, as seen in drones like the JOUAV CW-30E, a vertical take-off and landing (VTOL) craft. This drone integrates a combustion engine with electric propulsion, offering a maximum range of approximately two hundred kilometres, a top speed between eighty and ninety kilometres per hour, and a flight time of up to eight hours under ideal conditions.

However, such a machine comes with a significant price tag, often exceeding US$50,000, and its capabilities, while advanced, still fall short of some of the performance characteristics reported during these sightings.

Although there is no official specific noise emission data available for this drone, it is likely to be comparatively noisy, due to its size and hybrid gasoline propulsion system. Most of these nocturnal intrusions were reported to be silent.

Langley Air Force Base Incursions

At Langley Air Force Base in Virginia, there were seventeen consecutive nightly incursions of large, fixed wing drones of around six metres in length and travelling at up to one hundred and sixty kilometres per hour, or ninety nine miles per hour. Also reported were smaller, quad-copter type drones.

Six to eight drones penetrated restricted airspace approximately forty-five minutes after sunset until around midnight unchecked on a nightly basis, baffling military and government officials alike.

This particular series of events sparked an investigation by the All Domain Anomaly Resolution Office (AARO). Security personnel at the base reported these drones were unable to be tracked or intercepted and often became invisible to the naked eye.

Over the course of my interest, or near obsession with the UFO phenomenon, it was often cited that public witness testimonies could not be relied upon to be accurate, due to their amateur abilities at recognising aircraft and many other types of airborne elements.

I therefore find it difficult to believe that if these were aircraft, they were unable to be identified by professional military observers, or that highly skilled radar operators were unable to track them from, or to their landing point, considering the vast array of technologies available.

A simple google search regarding long-range drone specifications would surely narrow the search for a make or model if these flights were in fact authorised by the FAA, which was the final word on the matter from official channels. There wouldn't be many names on the list for US$50,000 drones, or for drones that are six metres long.

Sunset in Virginia in mid to late December is around 16:50. The objects arrived approximately 45 minutes later at 17:35. If we are conservative and say on average, they arrived at 18:00 each night, that is still over one hundred hours of constant sightings. They were unable to discover who was responsible for their incursion in the restricted air space of a USAF base.

Langley AFB serves as a headquarters for the Air Combat Command (ACC) and is home to the 1st Fighter Wing and the 363rd Intelligence, Surveillance and Reconnaissance Wing. There is also the reported speed of one hundred and sixty kilometres per hour, as this is almost twice the cruise speed of the VTOL JOUAV CW-30E which is reported to be ninety kilometres per hour, or fifty six miles per hour.

In the ~~UNITED STATES~~United States there are rules governing the engagement of unmanned aerial systems over domestic airspace, restricting any shooting down of a drone unless they posed an imminent threat. This situation, at a USAF base such as Langley, with unknown aerial vehicles making incursions over its airspace every night for seventeen consecutive nights, must have been considered a threat. Yet they were unable to track or stop them.

Drone Defence Technologies

Standard counter drone measures such as "Dronebusters" were reported as ineffective and unable to register some of the drones. A Dronebuster is a type of portable electronic counter Unmanned Aerial System (UAS) device, designed to disable drones without physically damaging them. It is a widely used, proven technology used by military and law enforcement.

Dronebusters are essentially portable handheld devices that flood the frequencies used by drones with RF noise. It disrupts the control link from the drone to the operator and affects the GPS signal, causing the drone to lose its position lock and depending on the drones programming, it will either return to its starting point, land immediately or just hover in place until the power source is exhausted.

This technology has an effective range of up to two kilometres, but the drawbacks of this system are, that it requires a line of sight to deliver the RF energy used to disrupt the drone. If the drone is controlled using frequencies outside of the operating parameters of the Dronebuster, it will be ineffective. There is also the possibility of autonomous drones using preprogrammed way points and which do not rely on GPS.

An enemy would be aware of this type of device and take countermeasures against it, but would an enemy have the technology to keep a drone of the size reported, around 6 metres, or nineteen feet eight inches, in the air for up to six hours and make it almost invisible to the naked eye.

The reported drones were described as belonging to unknown persons, were of an unknown type and specification, were unable to be tracked from their initial positions, or to their final destinations. They reached speeds almost double the speed which can be achieved in the commercial, long-range drone market, with flight endurance of six hours, plus travel to and from Langley AFB.

Although there are future possibilities for invisibility cloaking of the visible light spectrum, it is not currently possible to achieve this. With all the above information considered, official statements suggest that the FAA had approved these flights over a highly sensitive USAF base, does not add up.

Visible Light Spectrum Cloaking

The notion that these aerial objects might employ some form of invisibility or cloaking technology may seem far-fetched at first glance, yet theoretical and experimental advances suggest that such capabilities are not entirely out of reach.

There are several emerging technologies that could, at least in principle, contribute to reducing an object's visibility, whether in the infrared, radar, or even the visible spectrum.

One such method is known as optical cloaking. This cutting-edge approach relies on adaptive skin systems equipped with micro-cameras and micro-displays that work in tandem to capture the surrounding environment and project it in real time onto the surface of the object.

By effectively mirroring the background, the object becomes less discernible to the human eye, particularly when viewed from specific angles and distances. However, this technique does not provide perfect, 360-degree invisibility.

It is most effective when masking thermal signatures, making it potentially more useful in the infrared range than in visible light. Moreover, the complexity and energy demand of such systems are considerable, which may significantly reduce flight duration in airborne applications.

Another promising, though highly experimental, avenue involves meta-materials. These are artificially engineered materials designed to manipulate the behaviour of electromagnetic waves.

At present, meta-materials have demonstrated the ability to bend waves in the microwave and infrared ranges around an object, effectively guiding the waves so as to hide the object from detection.

Replicating this effect at the much shorter wavelengths of visible light remains a monumental challenge. While the underlying principles have been proven in controlled settings, practical implementation of full-spectrum optical invisibility using meta-materials remains elusive and is likely still decades away from deployment.

More grounded and perhaps more relevant in the context of unmanned aerial vehicles, is the use of radar-absorbing materials. These stealth technologies aim to minimise the radar cross-section of an object, rendering it less visible to radar detection systems. This approach has been employed in military aircraft and is increasingly being explored for UAV applications.

While still in development and not without limitations, this technology does offer a plausible explanation for the reduced radar visibility reported in some drone-related incidents. Together, these emerging methods of cloaking suggest that while true invisibility remains largely theoretical, partial stealth, particularly across certain wavelengths, may already be within reach, especially for highly specialised or covert platforms.

So where does this leave the drone hypothesis? The FAA flight approved drones explanation, although convenient enough to quell public concerns now that sightings have reduced in the US, it does fall short of a viable scenario.

There were many sightings in the ~~UNITED STATES~~United

States, and they can be easily researched, so I will refrain from documenting more at this time. From what we have seen, the phenomenon in the skies over the ~~UNITED STATES~~United States, cannot feasibly be attributed to what we would call conventional drones.

The inability of the United States military to be able to remove them from the air over sensitive military installations, the incredibly long flight times and also their ability to easily avoid detection hints that there are other options as to their identity.

Many descriptions of these objects report that although they looked like aircraft, many were silent and were not as large as they should be for the altitude they were at. I found these reports fascinating, as four or five months prior to the incidents in the ~~UNITED STATES~~United States, I had filmed what I believe to be an entity which seemed to mimic an aircraft.

Nearly a year after the initial reports, there are still sightings. The subject has been mostly dropped by mainstream media since the official FAA flight approved explanation.

European Drone Incursions

In September 2025, Denmark experienced similar incursions around Aalborg and Billund airports, and some military bases. These caused considerable disruptions and prevented flights from taking off. Danish authorities received over five hundred reports in a twenty four hour period. Some of these reports were explained away as "...just stars in the sky".

Following Denmark's incursions, other countries across Europe also had encounters. Poland, Germany and Norway have all had similar reports over airports, military bases, and sensitive infrastructure, once again causing flight delays and anxiety within governments and public alike.

This is reminiscent of the encounters in the US, however, the close proximity of Russia and the war with Ukraine, could very well alter NATO's perspective on the reported sightings. After reviewing some of the footage of the objects filmed in Europe, isolating images and clarifying the objects, I concluded that they are not drones.

17 Lights Over Fukushima

The Tōhoku Earthquake

On March 11th, 2011, at 14:46 JST (05:46 UTC), 70 km east of the Oshika peninsula, Honshu, Japan, the subduction of the Pacific tectonic plate beneath the Okhotsk microplate, often grouped with the North American tectonic plate, along the Japanese Trench, caused one of the most powerful earthquakes ever recorded globally.

It was certainly the most powerful recorded in Japanese history, with a magnitude of 9.0 to 9.1 and triggered a catastrophic tsunami with waves up to forty metres high. The waves travelled inland ten kilometres, flooding an area estimated to be over five-hundred and sixty square kilometres, also reaching coastlines across the Pacific.

Thirty minutes later, an aftershock of magnitude 7.9 was recorded, followed by hundreds more over the following weeks. Thousands of people were injured or displaced and almost twenty thousand people were confirmed dead or missing.

Mechanism of a Nuclear Disaster

As if this catastrophe was not enough, the Tōhoku Earthquake as it became known caused the nuclear reactors at the Fukushima Daiichi Nuclear Power Plant in Ōkuma, Fukushima Prefecture, to automatically shut down.

Backup diesel generators started, as designed, to power the cooling systems for the nuclear reactors but a fourteen metre high tsunami wave flooded the plant, disabling the generators and the electrical system.

Cooling of the reactors ceased, allowing reactors one, two and three to overheat which led to reactor core meltdowns. The consequent build-up of hydrogen gas caused explosions in buildings one, three and four, releasing significant amounts of radioactive material into the atmosphere and the Pacific Ocean.

The estimated total number of radioactive decays per second was in the order of one hundred quadrillion to one quintillion (10^{17} to 10^{18}). To place this measurement in context, if one becquerel is a single raindrop, a petabecquerel is a planet-wide deluge and Fukushima unleashed several hundred in one furious instant.

Measurement of Radiation and Half-life

The half-life of a radioactive isotope is the time it takes for half of its radioactive atoms to decay. This does not mean that all atoms will have completely decayed after two half-lives, or that radioactivity would cease. Instead, the remaining radioactive atoms are reduced by half again during each successive half-life.

This results in a diminishing but persistent level of radioactivity that can last for decades or even millennia. Radioactive releases are measured in becquerels (Bq), where 1 petabecquerel (PBq) equals one quadrillion (10^{15}) radioactive decays per second.

Major assessments estimate that Fukushima released hundreds of petabecquerels of short-lived iodine-131 and tens of petabecquerels of longer-lived caesium isotopes to the atmosphere and ocean.

Many zones within the plant now exhibit radiation levels low enough for workers to operate wearing standard protective gear, such as surgical masks and regular clothing; however, despite decontamination efforts, the interiors of the damaged reactor buildings, particularly units one, two, and three, are still showing dangerously high radiation levels.

Light Phenomena over Fukushima

With parallels to the New Jersey Drone incursions, where the majority of sightings were reported on the east coast, an area with a high density of nuclear power stations. In the weeks after the disaster, reports circulated of anomalous aerial objects seen over the plant, including accounts describing a metallic, disc-shaped object hovering above the site for an extended period before vanishing.

This incident was investigated by the Mutual UFO Network (MUFON) and covered by various news outlets, including The Huffington Post and The Independent. Chief Monk, Tomonori Izumi of Enmyoin Temple in Fukushima recounted witnessing many luminous UFOs following the nuclear explosion. He believed these UFOs came to readjust the flood of radioactive energy to protect the area. His account was featured in the Netflix docuseries Encounters, specifically in the episode titled "Lights Over Fukushima".

In response to the surge in UFO sightings, the town of Iinomachi in Fukushima Prefecture, known for its longstanding association with UFO phenomena, established the International UFO Laboratory in June 2021. Located near Mount Senganmori, a site with a history of reported UFO activity, the lab aims to collect, analyse, and publicise information on UFO sightings.

The facility has received hundreds of reports, some accompanied by photos and videos, which are evaluated for authenticity. When anomalous phenomena take interest in nuclear radiation, whether it be a nuclear power station, nuclear powered warships or a nuclear disaster like Fukushima, it is often cited that the phenomena are benevolent extraterrestrial beings, concerned that humanity will destroy Earth.

However, if we think in terms of the plasma entity hypothesis (PEH), it seems more likely their interest is possibly more self-serving. Maybe they are boosting their energy levels, healing or even exploiting a radiated environment for use in reproductive purposes.

At this time, we don't have enough information, but we can conclude from the last two chapters, that electromagnetic energy emissions and radiation, are an irresistible beacon to UAPs, whether they are extraterrestrial craft or intelligent, plasma-based lifeforms in the atmosphere.

Historical and Contemporary Comparisons

Across historical periods and cultural contexts, the phenomenon that humanity has interpreted as otherworldly has consistently presented itself as light. In the earliest recorded traditions, luminous beings are described as spheres, flames and radiant entities. In later centuries, the same motifs appear in accounts of apparitions, omens and celestial visitors.

In the modern era, the interpretations have shifted towards structured craft and metallic vehicles, yet the observational data remain heavily weighted toward sightings of luminous forms. If one examines testimony from both ancient and contemporary sources, it becomes evident that the primary feature reported is luminosity.

The interpretation of this luminosity depends largely upon the prevailing cosmology of the witnesses, but the underlying phenomenon appears unchanged. Ancient observers described the lights through the language and symbolism available to them.

Contemporary witnesses often attempt to frame similar luminous entities within technological narratives. What changes is the cultural framework, not the phenomenon itself. In the Book of Enoch, for example, luminous entities are described as "burning coals", "spheres of fire" and radiant forms that move with direction and intent.

The imagery is vivid and elemental, being neither mechanical nor inert, but living presences. Enoch reports visions in which spheres ascend and descend, appearing to breathe or pulse with inner energy. The descriptions emphasise movement, pattern and responsiveness rather than machinery or structure. These passages appear strikingly similar to many modern accounts from military observers stationed at nuclear facilities.

Contemporary testimonies, when stripped of interpretive

assumptions, reveal the same essential characteristics. At Malmstrom AFB in 1967, personnel described a glowing object hovering silently above the missile field. It appeared stable, self-luminous and capable of controlled movement.

At Ellsworth AFB, security officers reported a sphere of blue-white light that responded to their movements on the ground. At Kirtland AFB, guards witnessed a similar radiant orb that drifted above restricted areas with an apparent awareness of its surroundings. In Soviet accounts from the 1980s, missile officers also described red and orange spheres hovering above launch installations, pulsing rhythmically.

During the Fukushima disaster, cameras captured luminous globes drifting above the damaged reactors. Across these accounts, the witnesses emphasised the same qualities noted by ancient observers: luminosity: coherence, stability, responsiveness and directionality.

The following descriptions are from witnesses of reported phenomena over various nuclear missile silos and also Fukushima:

- *"...a glowing orb the size of a car"* – Kirtland AFB witness.
- *"...a sphere of blue fire hanging in the sky"* – Ellsworth AFB witness.
- *"...a brilliant light hovering silently"* – Malmstrom AFB witness.
- *"...a pulsating red globe above the silos"* – witness at Byelokoroviche, which is a former Soviet missile silo complex in, what is now Ukraine.
- *"...a ball of white light moving like a living thing"* – Fukushima

witness.

Viewed without the intervening filter of technological expectations, these testimonies speak less of structured craft and more of luminous intelligences. The consistent description of spheres, globes and fire-like forms suggests a phenomenon based on plasma-like behaviour rather than mechanical design.

Lights that move independently, hover without any obvious means of propulsion or aerodynamic surfaces and accelerate instantaneously, exhibit characteristics more closely aligned with organised energy than with physical vehicles.

This aligns coherently with the plasma entity hypothesis (PEH), which suggests that the intelligences observed across history may have bodies composed of charged, self-organising matter. The tendency to reinterpret luminous entities as craft appears to be a relatively recent development.

The twentieth century, shaped by aeronautics, rocketry and the Cold War, encouraged a technological framework through which anomalous observations were interpreted. The expression "Unidentified Flying Object" itself introduces a bias towards the assumption of a technological machine.

Witness accounts from the same period, however, continue to emphasise light, not structure. Many incidents, particularly around nuclear installations, involve descriptions of luminous orbs displaying controlled movement and apparent awareness. The craft interpretation often arises from cultural expectation, not from the observations themselves. If ancient observers attributed luminous intelligences to angels and divine messengers, and modern observers attribute them to extraterrestrial technology, the underlying assumption in both cases is that the beings originate from an external domain.

Yet the consistency of the phenomenon suggests that the intelligences may be neither divine nor extraterrestrial in the conventional sense. They may be indigenous to a plasma-based environment that surrounds and permeates the planet. They may represent a parallel ecology that interacts with human activity at critical thresholds, particularly those involving destabilising energies.

By incorporating this perspective, a continuity emerges between ancient accounts and modern reports. The watchers described in Enoch are not fundamentally different from the luminous objects reported over missile silos and nuclear reactors. Both eras of witnesses observed coherent light forms behaving purposefully, describing entities that appeared to communicate primarily through presence, pattern and motion.

Both interpreted the lights according to the dominant cognitive frameworks of their time. In this context, the nuclear-era sightings need not be viewed as evidence of extraterrestrial craft. Instead, they can be understood as encounters with luminous plasma intelligences responding to disturbances in the Earth's electromagnetic equilibrium.

The lights gather where energy is concentrated and where imbalance is greatest, where they appear in moments of profound transformation or danger. The presence of the phenomena at nuclear installations may represent a form of guardianship, maybe not in the mythological sense of divine intervention but more in an ecological sense of monitoring and maintaining balance within their own domain.

Thus, the luminous beings described throughout history may

form a single continuum. However, we must also consider the possibility of attraction to these imbalances is of a more mundane nature, possibly for energy harvesting or reproduction.

Whether witnessed by ancient prophets, medieval travellers or modern military personnel, the phenomenon is consistent. The lights may not be vessels; they may not necessarily be manufactured objects. The evidence points to radiant intelligences that appear to inhabit a hidden biosphere; a plasma-based environment interwoven with the physical world.

This perspective allows the present chapter to close with a unifying observation. The drones over New Jersey, the burning coals of Enoch, the orbs above missile fields and the spheres over Fukushima are not separate phenomena. They are iterations of the same enduring presence.

Human interpretation has shifted, but the phenomena have not. They materialise as radiant forms where there is an imbalance of energy, moving with intention. They are not defined by structured, manufactured craft or technology but by energy, field and form, remaining luminous and coherent.

The watchers of antiquity and the luminous structures of the nuclear age belong to the same lineage. Only our story surrounding the phenomena has changed.

18 Sky Trumpets as Atmospheric Voices

In previous chapters, we have discovered some of the diverse manifestations and abilities which may be attributed to plasma lifeforms, and that location, geology, culture, vocabulary and motive can affect humankind's perception of the phenomenon.

There are numerous extraordinary and mythological beings and strange occurrences described throughout human history, whose origins may well be founded in plasma-based life. Through diverse appearances and imaginative retellings, these accounts have taken on an evolution of their own.

These stories are often consigned to the realms of folklore, myth and fantasy, adding details which may not be accurate to the initial account with each reiteration. Humanity's appetite for such tales of legend and heroism seems limitless. This chapter seeks to introduce and explore some of these phenomena from the plasma entity perspective.

Sky Trumpets

Let us begin with one phenomenon which has still to be

definitively explained, yet has had little, if any mainstream media coverage. Nevertheless, this enigmatic subject, currently known as the *Sky Trumpet Phenomenon,* among other names, has been reported across continents and time, by diverse cultures.

I was mystified by this strange auditory occurrence, and realised that the proposed explanations really did not stack up. I came across reports of sky trumpets long before I began this research and became fascinated by the strange, disembodied sounds which reminded me of whale song. When we consider how cetaceans generate and use sound in water, there is a compelling analogy to be made.

Historical Reports

During the Roman era there were accounts of strange celestial sounds. One of the earliest known descriptions comes from Pliny the Elder (23 – 79 AD), the Roman author of the encyclopaedic *Naturalis Historia,* a thirty seven volume work that became foundational for ancient science and philosophy.

In "*Volume II: Astronomy and Meteorology*", Pliny describes bizarre noises emanating from the heavens, sounds similar to the clash of arms in the sky, which he attributed to natural yet mysterious phenomena. He used terms such as "*signa ex caelo*" or "signs from the sky", and "*clangores caeli*" or "sky clangors", and noted that these often preceded earthquakes or violent storms.

His writing suggests that these sounds were interpreted as divine signs or omens.

In the original Latin

"Audiuntur et caelestes clangores, armorumque fragores, quae non

tonitrua esse certum est. Haec auguria tempestatum aut terrae motus significant."

English translation

"There are also heard celestial clangors and the clashing of arms, which are certainly not thunder. These portend storms or movements of the earth."

There are also reports available from thousands of years before the Romans, in ancient China from the Zhou Dynasty, (~1000 to 256 BCE). Chinese historians documented unusual natural events with meticulous care. Chinese culture has had a long association with dragons, but what if their exalted dragons were in fact, another complex iteration of the luminous inhabitants of the plasma biosphere.

Works such as the "*Shujing*" or "Book of Documents", the "*Zuo Zhuan*" or "Commentary of Zuo", and later Han and Tang Dynasty records speak of "heavenly drums", "metallic sky sounds", and "silent lightning", all interpreted as celestial omens. These events were linked to impending plagues, natural disasters, or shifts in imperial power.

Native American Accounts

Among the Native American tribes of the Plains, particularly the Lakota, Dakota, Nakota, and Sioux, oral histories recount the presencc of "*Wakíŋyą̨n*", or "thunder beings". These powerful spiritual entities were associated with sky rumbles, metallic roars, and trumpet-like blasts heard in cloudless skies.

The Wakíŋyą̨n were seen as sacred messengers, their appearances signalling transformation, purification, or spiritual calling. These events were not always ominous. Instead, the thunder beings were believed to be both protectors and destroyers, forces of both renewal and reckoning. Strange sky sounds were sometimes interpreted as calls to embark on vision quests, or warnings to heed deeper spiritual truths.

Paracelsus and Sylphs of the Air

In the 16th century, Swiss German physician and alchemist Paracelsus (1493 – 1541) introduced a framework that attempted to reconcile mystical cosmology with natural philosophy. He proposed that each of the classical elements, Earth, Fire, Water, and Air, were inhabited by intelligent, invisible beings.

The air element, he claimed, was home to "*Sylphs*", aerial entities composed of air itself. In his "*Liber de Nymphis*" or "Book of Nymphs", Paracelsus wrote:

"The beings of the air are called Sylphs. They are mighty and have their bodies of the air. They are of human form and taller and stronger than humans. Yet they are invisible to human eyes, for their bodies are more subtle and transparent than the finest cloud."

Paracelsus continues:

"If the eyes of man were opened, he would see the people of the air, walking among the winds, shaping the clouds, and stirring the breezes... They guard the balance of vapours and the coursing of storms."

And most intriguingly:

"When the air trembles and the forests sigh without cause, when unseen hands pluck the strings of the wind, these are the Sylphs at play."

While Paracelsus does not directly reference the sky trumpet phenomenon, his descriptions are hauntingly consistent with modern interpretations; that intelligent, invisible beings shape and manipulate the air and its vibratory forces. What is particularly striking is that Roman, Chinese, Native American, and European Renaissance cultures, entirely isolated by geography and time, have all recorded similar sky phenomena.

None had communication with one another, yet all preserved traditions involving unexplained aerial sounds, believed to be signs or messages from powerful, non-human intelligences. This convergence lends strong support to the notion that the sky trumpet phenomenon is not a new or isolated anomaly, but rather a persistent and ancient aspect of human experience with the unknown.

Plasma Beings as Progenitors of Audible Frequencies

Let us now explore the possibility that plasma entities are the progenitors of the sky trumpet phenomenon. We have discussed many characteristics of these lifeforms in previous chapters, not least, their similarity to the biological cetaceans.

These mammals use sound in water very effectively, particularly for long-distance communication, as water conducts acoustic energy more efficiently than air. Some species, notably the humpback whale, have been known to have calls capable of traversing more than sixteen thousand kilometres or ten thousand miles.

Today, with access to sound recording technology and spectral analysis, we can study sky trumpet emissions in more detail. Recordings from around the world have shown that these sounds frequently fall within the 20 Hz to 150 Hz range.

Interestingly, frequencies below 20 Hz, which are classified as infrasound, are not audible to our ears but still have measurable effects on the human body. Infrasound can induce unease, nausea, anxiety, and a sensation of pressure, as well as cause physical objects to resonate. Structures such as buildings, bridges, and even entire landscapes may often vibrate sympathetically with these frequencies, generating secondary audible harmonics.

If, as is proposed, these beings are indeed responsible for the mysterious sky trumpet sounds, then it follows that they possess the ability to manipulate the air itself through a spectrum of resonant frequencies. These frequencies, based on numerous recordings and reports, appear to range from around 20 Hz, or possibly lower, up to about 150 Hz; 20 Hz being the lower limit of human hearing.

Plasma beings are theorised to vary tremendously in size, potentially up to many kilometres. Therefore, it may be possible to calculate the size of such entities responsible for these auditory events, using the sky trumpet frequency emissions to calculate the resonant cavity size. We can only speculate on whether the entity uses its whole form as a resonant cavity, or if the cavity is similar in size to a biological lifeform.

However, if we suppose that the resonant cavity size calculated is the minimum size of any frequency producing entity that produces the frequencies, then we can also scale up, based on known biological forms. This calculation is of course hypothetical, as there are many unknown elements and many variables. Maybe one day, far in the future, when plasma lifeforms are recognised and have been studied and documented by mainstream science, this line of investigation may be seen to be correct.

If so, it would be interesting to know the margin of error on the overall entity size. As a basis for these calculations, we will once again compare our plasma entities to their biological doppelgangers, the cetaceans, although adjustments will need to be made for the different mediums of water and air.

Whales communicate using frequency modulations specific to their environment, allowing sound to travel vast distances through the dense medium of water. If these sky-dwelling entities are similarly adapted to manipulate their own atmospheric medium, the parallels deepen.

Although we cannot yet determine how such beings might produce sound, given their semi-translucent, shape-shifting, plasma-like forms, it is conceivable that they modulate their density at will to interact with, or alter the surrounding environment.

Maybe some of these lifeforms may have achieved adept control over their own plasma densities within their semi-physical forms. A point to mention here is that, just as in a biological system, not all organisms are created equal. Due to the evolutionary process, many skills and abilities may be perfected by one species and discarded by another.

This has created a huge diversity of life on Earth and may well be paralleled in the hidden plasma biosphere. Over millennia, they may have learned not only to reshape, control, and disperse clouds but also to generate powerful vibratory frequencies capable of affecting both animate and inanimate objects.

If that is the case, the sounds we hear as sky trumpets could quite literally be their communications as vibrational expressions. In this sense, the phenomenon could be their voices.

Resonance: Open-Open, Closed-Open

An open – open resonance chamber is one in which both ends are free to oscillate. In such a system the fundamental mode forms a half-wave between the two open boundaries. This configuration is typical of cavities suspended freely in the atmosphere or within a symmetrical plasma structure, where no rigid surface constrains one end of the oscillation.

By contrast, an open – closed chamber is fixed or sealed at one end and open at the other, producing a quarter-wave resonance. This is the arrangement found in many biological systems such as cetaceans and humans, where the vocal tract terminates in a closed region at one end and an opening at the other.

The following images are not presented as evidence that the depicted structures are responsible for the sky trumpet phenomenon. Rather, they are included as illustrative examples of how large-scale atmospheric plasma structures, if present, could in principle support resonant modes consistent with reported acoustic characteristics.

In the images above, I have overlaid both the half-wave (open – open) and quarter-wave (open – closed) forms onto an image of an entity captured in the Pilbara region of Western Australia.

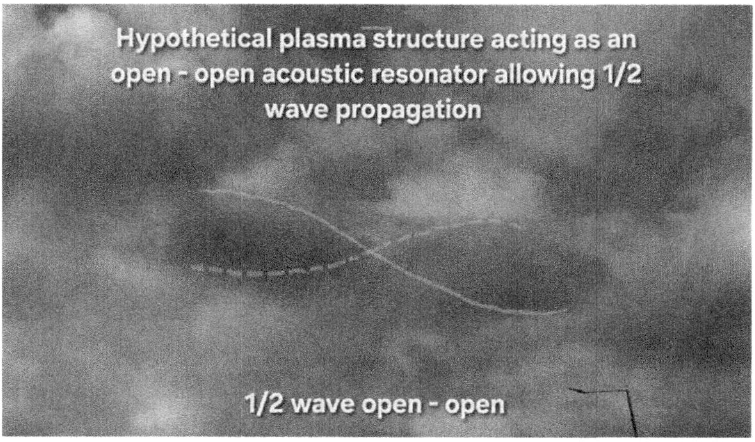

Fig. 5. Resonance mode: open - open.

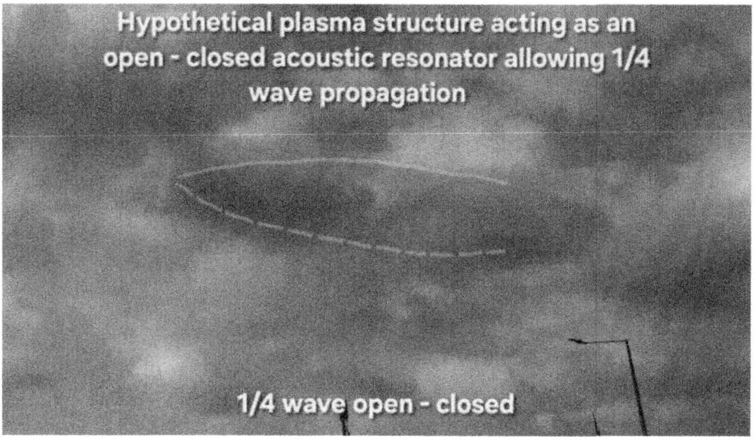

Fig. 6. Resonance mode: open - closed

These overlays illustrate how the same acoustic frequency can arise from different cavity geometries, and how the underlying boundary conditions determine the resonant behaviour of the entity.

Estimating Entity Size from Frequency

In this calculation, we will treat sky trumpet sources as quarter-wave, open-closed resonators, in the same family as biological vocal tracts in whales. For a given fundamental frequency f, the characteristic cavity length is taken as

$$L\_cav = v \div (4 \times f)$$

Where:

L_cav = the resonant cavity length

v = the local speed of sound in air (approximately $343\ m{\cdot}s^{-1}$ at 20° C)

f = the fundamental frequency

α = the proportional relationship between cavity length and characteristic cavity diameter, taken here as:

$$\alpha \approx 0.5$$

Here we take the characteristic cavity diameter as a fixed fraction of the cavity length, $D\_ent = \alpha \times L\_cav$, with $\alpha \approx 0.5$. Therefore, this waveform occupies a fixed fraction of this cavity length, such that

$$D\_ent = \alpha \times L\_cav$$

In this model, D_ent represents the characteristic diameter of the resonant cavity within the entity, rather than the full size of the organism itself. As with biological analogues such as cetaceans, the acoustic cavity occupies only a fraction of the overall body. Accordingly, the entity's true dimensions may well exceed the values calculated here; the model provides the scale of the sound-producing region, not its total extent.

This provides a simple, internally consistent way to estimate entity size from the observed acoustic frequency and to compare entities associated with different sky trumpet events on an equal footing.

A denser entity of the same size would be more massive and harder to drive dynamically, but it would not differ in its fundamental acoustic frequency, which is determined primarily by the geometry of the surrounding air column rather than by mass or density.

Using this framework, we can now apply the quarter-wave relationship to estimate plausible entity dimensions across the range of frequencies typically associated with reported sky trumpet events.

For a closed – open resonant system, the fundamental resonance condition is:

$$L\_cav = v \div (4 \times f)$$

The corresponding characteristic entity cavity diameter is then given by:

$$D\_ent = \alpha \times L\_cav$$

Substituting $\alpha \approx 0.5$ gives:

$$D\_ent = 0.5 \times (v \div (4 \times f))$$

Which simplifies to:

$$D\_ent = v \div (8 \times f)$$

Frequency (Hz)	Open-closed Cavity Length L_cav	Open-open Cavity Length L_cav	Open-closed Cavity Diameter D_ent	Open-open Cavity Diameter D_ent
5	17.15 m	34.30 m	8.57 m	17.14 m
8	10.72 m	21.44 m	5.36 m	10.72 m
10	8.57 m	17.14 m	4.29 m	8.58 m
12	7.15 m	14.3 m	3.57 m	7.14 m
15	5.72 m	11.44 m	2.86 m	5.72 m
20	4.29 m	8.58 m	2.14 m	4.28 m
25	3.43 m	6.86 m	1.72 m	3.44 m
30	2.86 m	5.72 m	1.43 m	2.86 m
40	2.14 m	4.28 m	1.07 m	2.14 m
50	1.72 m	3.44 m	0.86 m	1.72 m
60	1.43 m	2.86 m	0.71 m	1.42 m
80	1.07 m	2.14 m	0.54 m	1.08 m

Table 2. Estimated cavity lengths and characteristic cavity diameters for sky trumpet frequencies.

This provides a consistent way to translate audible frequencies into the physical scales of the plasma bodies hypothetically responsible for producing them. Lower frequencies correspond to longer resonant cavities and therefore to larger entities; higher frequencies imply smaller, more compact structures.

Using these relationships, the following calculated values illustrate the range of plausible entity cavity sizes, as shown in Table 2. Although these values are derived from a simplified acoustic model, they offer a coherent set of comparative scales. While many recordings cluster in the 20 Hz – 150 Hz audible range, some reported sky trumpet events exhibit dominant infrasonic components below 20 Hz, which are often inferred from secondary harmonics or physiological effects rather than direct audibility.

An entity producing a frequency of 5 Hz – 10 Hz, for example, would imply an approximate cavity diameter of four to nine metres, consistent with large, plasma lifeforms capable of sustaining organised field structures.

Conversely, sky trumpet frequencies in the 20 Hz – 40 Hz range correspond to resonance cavities in the one to two metre range, comparable to many of the luminous or structured forms recorded in both night-vision and visible-light observations.

Importantly, this approach allows events to be compared quantitatively. Two entities producing the same frequency may differ substantially in effective density or internal structure, but their acoustic size remains constrained by the shared requirement of driving a resonant cavity of a particular length.

In this way, the sky trumpet phenomenon offers a rare opportunity to infer physical dimensions of otherwise non-solid or transient plasma-based lifeforms through their interaction with the surrounding atmosphere.

A brief caveat is warranted: the calculations above assume a closed – open configuration, which is appropriate for phenomena occurring close to the ground or near strong atmospheric gradients.

Events occurring at higher altitude, where the entity is suspended freely with no nearby reflective boundary, may instead behave as open – open resonators. Such cases would require a simple modification of the model, effectively doubling the inferred cavity lengths and thus the characteristic cavity diameters, as is shown in Table 2.

Nevertheless, for the majority of reported sky trumpet events, particularly those recorded at ground level, the closed – open approximation appears to be the most physically appropriate. In toothed whales, such as orca, sound is generated in a complex system of nasal sacs and phonic lips located just below the blowhole, rather than along the entire body.

Detailed measurements of the full vocal tract length in orca are not available in the literature. However, based on skull proportions and the localisation of the sound producing apparatus in the head, the effective resonance cavity is likely to occupy only a small fraction of the animal's total length, perhaps on the order of ten to thirty percent.

In a six to nine metre orca, this would correspond to a resonant path of roughly one, to just a few metres. By analogy, the cavity sizes inferred from sky trumpet frequencies should be interpreted as the scale of the acoustically active region of a plasma entity, not its full spatial extent.

As with orca, the organism could be significantly larger than the cavity that actually drives the resonance. If we adopt a conservative biological analogy and assume that the resonant cavity occupies on average around twenty per cent of an entity's characteristic diameter, the closed–open values in Table 2 imply entities ranging from tens of metres down to only a few metres across.

At the lowest frequencies (5 Hz to 8 Hz), cavity diameters of five to nine metres correspond to entities in the approximate twenty five metre to forty metre range, consistent with large, slow-moving atmospheric plasma lifeforms.

Frequency	Cavity Diameter D_CAV	1/4 Wavelength Entity Length	1/2 Wavelength Entity Length
(Hz)	(m)	(m)	(m)
5	8.60	43.0	86.0
8	5.40	27.0	54.0
10	4.30	21.0	42.0
12	3.60	18.0	36.0
15	2.90	14.0	28.0
20	2.10	11.0	22.0
25	1.70	8.6	17.2
30	1.40	7.2	14.4
40	1.10	5.4	10.8
50	0.86	4.3	8.6
60	0.71	3.6	7.2
80	0.54	2.7	5.4

Table 3. Estimated entity lengths based on frequency for ¼ & ½ wavelengths.

Frequencies between about 10 Hz and 20 Hz yield entity diameters of roughly ten to twenty metres, while the more commonly reported trumpet bands between 20 Hz and 40 Hz correspond to entities in the five to ten metre range. At still higher frequencies (50 Hz – 80 Hz), the inferred entity size falls to only three or four metres.

These estimates are of course highly speculative, but they demonstrate that even under conservative assumptions the sky trumpet phenomenon is compatible with a population of plasma lifeforms spanning from a relatively compact few-metres scale to multi tens of metres. Should a particular event be better described by an open – open resonance, the same frequencies would imply entities approximately twice as large again. These scaled entity sizes for quarter and half wavelengths can be seen in Table 3.

Modern Case Studies

Kiev, Ukraine (2011)

In August 2011, residents across Kiev recorded a series of metallic, trumpet-like resonances that persisted for several minutes and were audible across multiple suburbs. Independent recordings captured dominant frequency bands between 18 Hz – 25 Hz, with secondary harmonics emerging around 40 Hz – 50 Hz, as confirmed by spectral analyses conducted by amateur acousticians.

Proposed explanations, ranging from industrial machinery to atmospheric ducting, failed to account for the event's persistence, harmonic stability, and city-wide audibility. Notably, the primary frequency range aligns closely with the cavity diameters estimated earlier in this chapter, implying a resonant source of several metres in scale. While no definitive cause was identified, the Kiev event remains one of the clearest modern analogues to the ancient sky trumpet phenomenon.

British Columbia, Canada (2020)

In May 2020, multiple communities in British Columbia reported a deep, metallic sky sound lasting nearly 10 minutes, described variously as "*a giant horn,*" "*a metallic bow being drawn across the sky,*" and notably, "*a whale song in the clouds.*" Witnesses reported the absence of aircraft or industrial activity, and local authorities confirmed no construction, seismic blasting, or atmospheric research operations in the region at the time.

Video recordings revealed a dominant frequency of approximately 14 Hz – 16 Hz, with faint higher-order harmonics. This falls within the lower band of sky trumpet cases and corresponds, according to the resonance model outlined earlier, to cavity diameters between roughly ten and twelve metres for a closed – open system.

The event's isolated nature and lack of mechanical signatures challenge purely environmental explanations and support the possibility of a natural, self-organising resonant source in the atmosphere.

Slovakia (2016)

A widely circulated 2016 incident recorded in several towns across Slovakia captured a series of loud, resonant blasts that appeared to emanate from the sky rather than the ground. Witnesses described the sounds as "*metallic roars*" and "*trumpet-like waves*" lasting several minutes. Acoustic analysis of uploaded videos showed primary peaks around 20 Hz – 22 Hz with consistent amplitude modulation, indicative of an oscillatory source rather than a random atmospheric rumble.

Meteorological services proposed explanations such as temperature inversions or shifting air masses, but these did not convincingly account for the harmonic structure or the rhythmic periodicity evident in the recordings. When interpreted through the plasma-entity hypothesis, the event could represent a mid-range resonator with a cavity diameter of seven to eight metres, producing coherent, low-frequency emissions detectable over long distances.

Conclusion

The sky trumpet phenomenon is ancient, well-documented and globally consistent across cultures and centuries. While science seeks naturalistic explanations such as shifting tectonic plates and upper atmospheric jet streams, the ethno-historical and metaphysical lens offers an equally compelling narrative.

A viable case can be made that intelligent, plasma-based lifeforms have existed in our atmosphere since the beginning of human history, perhaps long before. If so, these unidentified atmospheric resonances may be yet another manifestation of the multifaceted hidden biosphere.

Perhaps we are not merely registering sound, but encountering a language we have not yet learned to comprehend, an expression of communication from a form of intelligence still unrecognised by science.

If these signals contain meaning, what might they reveal once we understand them? And more provocatively: how much have we failed to notice until now? In the next chapter, we turn to a physical phenomenon debated for decades, one that may offer further clues to the presence of atmospheric plasma intelligences. That phenomenon is known as crop circles.

19 Crop Circles

A Plasma-Ethological Interpretation

For more than four decades, crop circles have existed in a liminal space between natural anomaly, artistic intervention, and alleged extraterrestrial communication. Debate has often polarised around these categories, with arguments oscillating between mechanical hoaxes, meteorological curiosities, and symbolic messages from elsewhere.

Yet this framing, compelling as it may be, risks obscuring a more fundamental possibility; that the phenomenon is neither symbolic nor fraudulent, but ethological in nature. When examined through the plasma entity hypothesis (PEH), crop circles cease to resemble encoded diagrams or deceptive constructions and instead emerge as behavioural artefacts, physical imprints produced when atmospheric plasma lifeforms interact with the terrestrial environment.

In this perspective, the patterns recorded in fields are not communications but consequences: the geometric residues of movement, resonance, and engagement by coherent, intelligent plasma structures. The purpose of this chapter is not to retell the well-worn history of crop formations or to adjudicate human involvement.

Rather, it is to reinterpret the phenomenon through the combined lenses of ethology, plasma physics, and atmospheric dynamics. The central proposition is simple yet far-reaching: Crop formations may represent visible traces of behaviour, expressions of curiosity, interaction, or even play, performed by non-biological yet intelligent atmospheric organisms.

A Shift in Perspective

Attempts to interpret crop circles as encoded messages or deliberate symbolic communications have always struggled with the absence of any coherent linguistic structure. The formations themselves are elegant, precise, and often mathematically resonant, yet they do not exhibit the repeatability or internal grammar one would expect from an intentional sign system.

When considered through the lens of the plasma entity hypothesis (PEH), however, their complexity no longer appears as a linguistic puzzle awaiting translation, but as an ethological clue, an artefact of behaviour rather than language. In biological systems, intelligence frequently expresses itself not through symbols but through movement, pattern, and interaction with the surrounding medium.

Dolphins, for example, generate toroidal vortices, perfectly formed bubble rings, purely for social engagement or exploration. These rings, transient yet highly structured, arise from the animals' mastery of fluid dynamics rather than from any attempt to communicate abstract ideas. Starlings in murmuration produce vast, shape-shifting geometries, coordinated not by central command but by local, responsive interactions, creating patterns that seem almost mathematical in their coherence.

Schools of fish generate spirals, waves, and interference fringes as they reorganise themselves within a shared flow field to avoid predation, while flocks of birds inscribe fleeting aerial lattices that form and dissolve without leaving any message beyond the choreography of the moment.

These natural behaviours share a set of defining characteristics; they are fluid, emergent, rhythmic, and often aesthetically compelling, but they are not encoded narratives. They are the visible consequence of living systems interacting with their environment, behaviour recorded for an instant in the medium through which the organism moves.

When crop formations are approached with this model in mind, they begin to resemble a similar class of patterned trace; the imprint left when an intelligent, dynamic organism, if plasma entities may indeed be thought of as organisms, moves, rotates, oscillates, or interacts with the atmospheric boundary layer just above a field. From this perspective, the geometric precision of crop circles is not evidence of deliberate symbolic communication but the physical memory of a behavioural event.

The geometry arises not from intention to convey meaning to humans, but from the intrinsic order of field dynamics and movement patterns within plasma-based lifeforms. Complexity becomes a signature of behaviour, interaction, or even play; the field imprint becomes analogous to a dolphin's bubble ring or a murmuration's transient architecture.

Under the PEH framework, crop circles can therefore be reinterpreted as behavioural residues, momentary expressions of plasma entities interacting with the terrestrial environment, preserved because their medium happened to be a malleable one, the compliant stems of cereal crops.

Expressions of Plasma Intelligence

If plasma entities possess structured electromagnetic architectures, as argued in earlier chapters, then their movement through the lower atmosphere would naturally generate coherent field geometries. These geometries would not be static; rather, they would be dynamic expressions of oscillation, rotation, and internal field modulation, shaped by the entity's morphology and the surrounding environmental conditions.

When such a structured field intersects with compliant biological material, wheat, barley, rapeseed, or any crop capable of bending without immediately breaking, the result is not a message inscribed with intention, but an imprint; a physical memory of the interaction between a coherent plasma form and the terrestrial medium it momentarily touched.

Under this interpretation, crop circles become field-mediated interactions rather than deliberate constructions. They are expressive behaviours in the same way a dolphin's vortex ring or a murmuration's shifting lattice is expressive, not communicative in the linguistic sense, but indicative of an organism's structure, motion, and internal coherence.

The patterns themselves serve as structural echoes of plasma motion, a kind of harmonic residue recording the dynamic interplay between oscillating electromagnetic fields and the plants' cellular architecture. In this context, complexity emerges through resonance rather than through symbolic intention.

Interference patterns, spirals, nested rings, and even the occasional fractal-like geometry may reflect the superposition of multiple field modes or the brief synchronisation of different oscillatory components within the entity. The physical effects consistently observed in genuine formations lend further support to this interpretation.

Expansion of stem nodes, sometimes accompanied by evidence of localised heating, suggests exposure to short-duration energy pulses rather than prolonged mechanical pressure. Microwave-like signatures have been reported in several examined circles, consistent with rapid dielectric heating capable of softening plant tissue just long enough for stems to bend without fracturing.

The intricate layering of fallen crop, stems woven into multi-directional arrangements, overlain in sequences impossible to achieve by simple mechanical flattening, points to a process that acts volumetrically rather than through surface contact. Such characteristics align more coherently with a plasma-driven mechanism than with either human hoaxing or extraterrestrial technology.

A human group working mechanically cannot produce node expansion or uniform microwave effects; a hypothetical technological device would likely leave signatures of propulsion, radiation leakage, or directional force. Plasma entities, by contrast, are hypothesised to interact with matter through precisely the kind of non-contact electromagnetic processes observed in genuine formations.

Their structured fields, resonant dynamics, and capacity to modulate energy locally would explain both the precision of the geometry and the biological anomalies within the plants.

Thus, when crop circles are reframed as the physical memory of plasma behaviour, a trace of movement, resonance, or interaction with the boundary layer, and not as intentional messages, the phenomenon becomes neither supernatural nor inscrutable.

It becomes biological in the broadest sense; the patterned footprint of a lifeform whose form finds structure in energetic architecture.

Evolution of Complexity

One of the most striking characteristics of the crop circle phenomenon is the marked escalation in complexity since the late twentieth century. Simple circular forms, once the dominant type, were gradually followed by intricate geometric arrays, multi-layered structures, fractalised motifs, and formations exhibiting harmonic or mathematical sophistication.

While this progression is often attributed to human ingenuity, the biophysical anomalies present in many genuine formations, combined with the extraordinary precision and speed with which they appear, suggest the influence of a process more subtle and more intelligent than mechanical flattening. Within the plasma entity hypothesis (PEH), increasing complexity is not viewed as an artistic evolution but as the natural outcome of biological, environmental, and behavioural dynamics.

If plasma-based lifeforms undergo developmental stages or energetic maturation, their internal resonance geometries would be expected to grow more complex over time. In this view, the simplest circles may represent the signatures of smaller, younger, or less stable organisms, entities whose electromagnetic architecture has not yet developed the full richness of harmonic expression.

Conversely, larger or more mature plasma beings, capable of sustaining intricate internal field interactions, would naturally produce multi-ring structures, nested harmonics, interlocking lattices, and complex rotational patterns when interacting with compliant terrestrial surfaces.

This concept mirrors biological precedents; juvenile whales vocalise in relatively simple sequences compared to the elaborate, evolving songs of mature individuals; fledgling birds exhibit rudimentary displays long before the full courtship performances of adulthood.

Complexity, in this sense, is a function of developmental trajectory rather than communicative intention. Environmental changes may also play a decisive role. The electromagnetic character of Earth's atmosphere has transformed dramatically in recent decades.

High-voltage transmission lines, radar arrays, telecommunications towers, hyperscale data infrastructure, and proliferating satellite constellations have created a denser and more variable electromagnetic environment than at any previous point in recorded human history.

If plasma entities are highly sensitive to such fields, as suggested by their behaviour around electrical storms, volcanic lightning, nuclear reactors, and other energetic environments, then increased ambient energy availability could enable more stable, coherent, or powerful expressions of their field dynamics.

In this context, the rising complexity of crop formations may reflect not an evolution of intention, but an evolution of conditions: the atmospheric medium itself now provides more structure, more energy, and more resonance channels through which a plasma organism might interact with the physical world.

Perhaps the most compelling possibility, however, lies in ethology. Many intelligent species alter their behaviour when they become aware of being observed. Dolphins produce elaborate bubble rings and perform acrobatic manoeuvres around boats; elephants revisit human encampments when something captures their interest; corvids modify their tool use in response to human problem-solving.

Awareness of attention, whether visual, auditory, or emotional, can elicit play, experimentation, or display. If plasma entities perceive human presence not through vision but through shifts in electromagnetic coherence, our cameras, drones, devices, and even the synchronised cognitive focus of gathered observers, then their behaviour may change accordingly.

The elaboration of crop formations could, under this interpretation, be a feedback phenomenon; a dynamic interplay between human curiosity and plasma entity response, where complexity arises from a reciprocal interaction rather than one-sided production.

In this light, the crop circle phenomenon may be understood not as an attempt at communication in a symbolic sense, nor as mere environmental disturbance, but as the expressive behaviour of an intelligent plasma-based organism responding to developmental changes, environmental stimulation, and interspecies attention. Complexity, therefore, is neither accidental nor solely artistic; it is ecological, ethological, and emergent.

Group Behaviour and Field Dynamics

The most elaborate crop formations often display degrees of symmetry, harmonic layering, and rotational interference that exceed what a single coherent field is likely to produce. Their internal geometries appear to arise not from a solitary agent but from the intersection of multiple dynamic forces, each contributing a distinct oscillatory signature.

When viewed through the plasma entity hypothesis (PEH), these complex structures may reflect episodes of group behaviour among plasma lifeforms, communal interactions whose energetic overlap leaves patterned traces in the environment.

In biological systems, complexity frequently emerges from coordinated activity rather than individual action. Murmurations of starlings generate sweeping, fractal-like geometries that shift in microseconds, revealing the capacity of distributed organisms to produce coherent, large-scale patterns without central control.

Dolphins, when hunting cooperatively, create spiral or toroidal formations of bubbles that corral fish with remarkable spatial precision. Even the synchronised bioluminescent pulses of certain marine species produce emergent wavefronts of light across vast stretches of ocean.

These phenomena are not forms of symbolic communication directed at other species; rather, they are spontaneous expressions of collective behaviour, shaped by shared sensory fields and responsive feedback loops. If plasma entities interact socially or communally, similar dynamics may be at work.

Their electromagnetic fields, each one a structured, oscillating envelope, would naturally interfere, overlap, and resonate when multiple entities occupy the same space. The resulting interference patterns could be extraordinarily complex. Standing waves, rotational harmonics, or spiralling gradients produced by the mutual modulation of neighbouring field structures, rather than by deliberate design.

When such cooperative interactions occur above a compliant medium such as wheat or barley, the collective dynamics may be imprinted physically as a coherent pattern, a momentary record of plasma sociality.

Under this interpretation, the largest and most intricate formations are not messages, warnings, or attempts at symbolic communication. They are behavioural artefacts, the atmospheric equivalent of a murmuration's geometry or a dolphin pod's coordinated spirals, expressions of group dynamics rather than linguistic intention.

The field patterns preserved in crops would thus represent a rare insight into the communal life of these plasma organisms; how they cluster, interact, stabilise, or perform collective displays. In this context, complexity becomes evidence of social structure, not artistry of interaction nor instruction.

Movement through a medium always leaves traces. Water records the vortices and rolling wakes of creatures passing through it; air retains, however briefly, the contrails, pressure waves, and wing-tip vortices produced by birds or aircraft. Each medium registers motion according to its own physical laws, preserving momentary signatures of the forces that have travelled through it.

If plasma entities move through the lower atmosphere in a similarly fluid manner, particularly during periods of localised temperature, humidity, or electromagnetic contrast, their passage would likewise inscribe characteristic patterns upon the environment.

A plasma organism travelling close to the ground would be expected to generate toroidal shear structures, concentric pressure gradients, and rotational fields arising from the interaction of its own electromagnetic architecture with the surrounding air.

The release of microwave-level energy, whether as a by-product of locomotion, stabilisation, or field reconfiguration, could soften or elongate plant stem nodes, while swirling boundary-layer dynamics might create the distinctive rotational layering seen within many formations.

None of these signatures need imply intention. Rather, they are the atmospheric equivalents of wakes and eddies, the physical consequences of a body in motion. In this sense, crop circles may be understood as plasma "swim patterns", kinematic memories of movement, captured temporarily by the compliant medium of wheat or barley.

The formations would then serve as residual impressions of the entity's passage, shaped not by symbolic design but by the interplay of motion, field dynamics, and environmental conditions.

What appears to the human eye as an intentional glyph may thus be analogous to the spiralling trails left by dolphins as they arc through water or the vortical patterns left by birds navigating complex air currents. The crop circle becomes less a message and more a moment of interaction; a physical record of plasma locomotion preserved in the landscape.

THE QUESTION OF COMMUNICATION

A persistent assumption in popular discourse is that crop circles constitute messages, deliberate attempts at communication directed toward humanity. Yet meaningful communication requires far more than pattern or beauty alone.

It demands a repeatable syntax, a stable set of semantic rules, and clear evidence of intentional encoding for a recipient capable of comprehension. No known crop circle formation, however elaborate, satisfies these criteria. The patterns do not repeat with linguistic regularity, nor do they exhibit any internal grammar or decipherable system of signs.

Instead, genuine formations display qualities that belong not to language but to behaviour: fluid improvisation, mathematical resonance, harmonic symmetry, and a responsiveness that sometimes seems attuned to environmental conditions or even to human presence.

These characteristics align less with symbolic communication and more with the expressive repertoires observed in many intelligent species. Dolphins create bubble rings not to transmit information but as hunting techniques, acts of play, and social bonding. Birds engage in elaborate aerial displays that are aesthetically complex yet not linguistic.

Bioluminescent marine organisms produce coordinated patterns that are emergent rather than encoded. Such behaviours are meaningful within the species' own perceptual world but are not designed to be read by outsiders. Under the plasma entity hypothesis (PEH), crop circles may be understood in a similar way, as expressive acts without the expectation of interpretation.

They could hold aesthetic, structural, or energetic significance within the entities' own context, arising naturally from their motion, their internal field dynamics, or their interactions with one another. This need not imply that the formations are intended as messages for humanity.

What we witness may instead be the incidental traces of movement, the atmospheric equivalent of a dance or display that is meaningful to the performer but only partially perceptible to the observer.

In this framing, the question shifts from, "What are they trying to tell us?", to the more revealing, "What behaviour are we witnessing?" The answer may lie not in decoding symbols but in recognising that we are observing a form of life whose expressive medium is not only sound, as in the sky trumpet phenomenon, but also geometry, inscribed briefly upon the landscape.

Crucially, this interpretation does not preclude alternative explanations, but it reframes the phenomenon in a way that generates testable expectations: non-contact energy signatures, consistent node deformation, and field-coherent geometries independent of human construction methods.

A Plasma-Ethological Model of Formation

Within the Plasma Entity Hypothesis, the creation of a crop circle can be understood not as an act of inscription, but as the natural outcome of an interaction between plasma lifeforms and the atmospheric–biological medium through which it moves.

A plausible sequence begins with a stimulus, a shift in atmospheric stability, a charged inversion layer, an influx of electromagnetic activity, or even the presence of humans whose technologies subtly alter local field conditions.

Such perturbations may attract or activate nearby plasma entities, much as fluctuations in ocean currents draw marine animals or sudden thermals elicit responses from birds. Once engaged, the entity's behaviour becomes the primary driver of pattern formation.

Plasma lifeforms, as theorised in previous chapters, possess the ability to modulate their density, rotate around internal fractal axes, and engage in vibrational or oscillatory modes of movement.

These behaviours, analogous to the dynamic locomotion of dolphins, the spiralling dances of starlings, or the vortex play of whales, would naturally generate coherent electromagnetic fields that extend beyond the boundaries of the entity's visible form.

From this behavioural and energetic activity arises the projection of stable or semi-stable field geometries into the surrounding environment. The entity's internal resonances, rotational symmetries, and oscillatory harmonics can produce standing-wave configurations, toroidal, helical, or fractal in nature, that momentarily structure the air and electromagnetic gradients around them.

These fields, when they intersect with a compliant medium such as wheat or barley, initiate a process of coupling. Plant stems soften, nodes expand under brief heating, and the crop bends without breakage, consistent with observations documented in genuine formations.

Such microwave-like effects, localised thermal anomalies, and complex multi-layered lays of crop, are all compatible with short-lived but intense energetic interactions rather than mechanical force.

The final stage is the emergence of the pattern itself, not as a symbol deliberately crafted for human interpretation, but as the geometric shadow of an organism's movement and internal field structure.

The resulting formation is therefore an imprint; a transient memory of behaviour preserved in the biological substrate. In this model, crop circles require no linguistic intent, no communicative motive, and no external technology.

They arise instead from the interplay of behaviour, field dynamics, and environmental responsiveness, forming a natural expression of plasma ethology rather than a message awaiting decipherment.

Conclusion

When freed from the restrictive binaries of human hoax or extraterrestrial message, crop circles begin to reveal a more nuanced and compelling possibility. They are physical traces left by intelligent, coherent, plasma-based lifeforms interacting with the terrestrial environment. Interpreted through the plasma entity hypothesis (PEH), the formations shift from encoded communications to ethological signatures, expressive, dynamic, and responsive.

Their increasing complexity over recent decades need not imply an escalating attempt to communicate, nor a secret symbolic language. Instead, it may reflect the maturation of plasma organisms, the influence of a technologically saturated electromagnetic environment, or the emergence of interspecies curiosity as plasma entities become aware of human attention and presence.

What appears as deliberate design may, in this context, be better understood as a form of behavioural expression, analogous to the vortical air bubbles of dolphins, or the fluid geometries produced by murmuring starlings. Under this reinterpretation, crop circles are not messages to be decoded but geometric residues of behaviour, ephemeral interfaces between two very different ecologies.

They are moments in which the hidden biosphere intersects briefly with the visible world, leaving behind patterns created not through mechanical force but through field dynamics, resonance, and movement. The flattened wheat becomes a record of encounter, a canvas on which another form of intelligence has inadvertently inscribed its passage.

In the broader context of this book, crop circles may represent one of the earliest and most accessible crossover phenomena between species separated only by spectrum. They invite us to rethink what communication might look like when it arises from forms of life whose biology is electrical, whose behaviours are harmonic, and whose presence becomes visible when interacting with matter dense enough to record their passing.

Thus, the question is no longer, "What kind of beings create them?", instead "What has changed that this mode of expression is now leaving such visible traces?" The formations scattered across fields in southern England and around the world, may be far more than aesthetic curiosities. They may be the first steps in recognising that humanity is not an intelligence alone, but shares Earth with intelligent inhabitants of a hidden biosphere. An atmospheric ecology that has been ever present but rarely perceived.

20 Dragons and Sky Serpents

Reinterpreting the Dragon Myth

Across the world's oldest cultures, long before scientific frameworks existed to categorise natural phenomena, humans described encounters with radiant, serpentine beings that moved through the sky with intelligence, with agency, and often with startling power.

These entities appeared in forms that were luminous, elongated, shapeshifting, sometimes winged but often wingless, and almost always associated with storms, fire, water, and celestial disturbance. Their universality has led many modern scholars to treat them as archetypal expressions of the collective mind.

Once again, a different interpretation becomes possible when viewed through the plasma entity hypothesis (PEH). The dragon may not be a symbolic construct, nor a purely psychological artefact, but a cultural record of encounters with plasma-based lifeforms whose energetic properties, behaviour, and form were faithfully observed but imperfectly understood.

This interpretation becomes particularly compelling when we recognise that ancient observers were direct witnesses to atmospheric phenomena that modern plasma physics only began to describe in the twentieth century.

The mismatch between observation and vocabulary meant that pre-scientific cultures could only interpret such encounters through the symbolic language available to them, leading to remarkably consistent descriptions of luminous, serpentine intelligences long before such forms were understood as natural plasma configurations.

Ancient linguistic clues already point in this direction. The Greek "*drakōn*" derives from *derkomai*, meaning "to see clearly" or "to flash", emphasising brilliance and sudden appearance rather than biological anatomy.

Chinese descriptions consistently refer to coiling light, radiant bodies, cloud-riding serpents, and creatures whose breath shines like fire. Norse sagas record serpents of flame twisting through the heavens, and early British chronicles describe "fiery flying serpents" taken as omens of disaster or change.

These accounts, scattered across continents with no possibility of direct cultural transmission, depict the same essential form; a luminous, flexible, sinuous and aerial presence behaving in ways that align more closely with plasma physics than with zoology.

Their global consistency suggests not a shared mythology, but a shared stimulus; a recurring class of aerial phenomenon capable of appearing wherever atmospheric conditions favoured plasma formation.

Serpentine Form as Plasma Morphology

Plasma, as the dominant state of matter in the universe, rarely forms static, spherical bodies. Its natural shapes are filamentary and serpentine, shaped and confined by magnetic fields into twisting ropes, braids, helices, and writhing cords of light.

Birkeland currents in the Earth's magnetosphere demonstrate this vividly, coiling and turning like immense cosmic serpents. In laboratories, dusty plasma behaves similarly, producing undulating, serpentine filaments capable of self-organisation and quasi-lifelike motion.

Even atmospheric plasma events, sprites, jets, ELVES, and blue jets, regularly manifest as darting, branching, or writhing columns of light that observers describe as snake-like.

Modern high-speed imaging has confirmed that these plasma structures can persist for seconds, exhibit internal organisation, and propagate along electromagnetic filaments, traits that ancient cultures recorded as movement, intent, and even emotion.

To ancient eyes, deprived of the conceptual vocabulary required to describe charged-particle dynamics, these phenomena naturally resembled serpents.

The absence of wings in many mythic dragons becomes an important clue; a wingless creature capable of coherent flight contradicts biological expectation but aligns perfectly with a plasma-based organism, whose manoeuvrability would derive from electromagnetic coupling rather than aerodynamics.

Indeed, in my own field recordings, elongated objects, hypothesised to be plasma entities, have appeared to ripple and undulate through the atmosphere as if with purpose.

This controlled, fluid-like behaviour is remarkably consistent with the serpentine motion described in ancient myth. The parallel between ancient descriptions and modern footage becomes increasingly difficult to attribute to coincidence when both display the same morphological trait; locomotion without physical propulsion, achieved instead through field modulation.

The Fire Breathing Problem

One of the most persistent features of dragon lore is the ability to emit or breathe fire, a trait often dismissed as fanciful. Yet ancient texts rarely describe combustion in a biological sense; instead, they speak of radiant exhalations, bright flares, luminous vapour, or glowing streams of light issuing from the creature's mouth or core.

Plasma entities undergoing rapid energy reconfiguration may emit incandescent jets, coronal discharges, and photonic bursts that might appear as fire to the naked eye. Once again, this is likely a function of available vocabulary. In some cases, this may involve a sudden recombination of charge within a resonance chamber-like structure, precisely the glowing "pearl" depicted so frequently in Eastern iconography.

This interpretation is strengthened by documented cases of ball lightning emitting streamers, jets, and pulsed bursts of energy that observers routinely describe as "fire".

Ancient witnesses, lacking a framework for electrical discharge phenomena, naturally framed these radiative events as "fire-breathing." Observations have repeatedly shown entities whose "mouths", or anterior regions, brighten intensely during moments of manoeuvre or structural reconfiguration, and then fade.

These radiative bursts may sometimes be brief yet comparatively bright, and might well have appeared to ancient witnesses exactly as the fire-breathing attributes of dragons. Thus, the fire-breathing fable may be plasma physics interpreted through pre-scientific sensory frameworks.

Dragons as Weather and Energy Beings

Nearly all dragon traditions associate these beings with storms, winds, rains, floods, lightning, and profound atmospheric disturbance. This is not poetic embellishment; it reflects plasma's natural tendency to appear in regions of atmospheric instability.

High electrical potential charged cloud layers, solar-induced geomagnetic disturbance, or volcanic ash are known to support plasma formation and sustain aerial plasma structures. In many of my observations, entities appear consistently during thunderstorms, willy-willies, and other high-energy environmental conditions, behaviour that directly parallels the mythic dragon's role as storm-bringer or weather-controller.

Plasma researchers have long recognised that atmospheric plasmas can reach considerable scales during such events, sometimes forming structures hundreds of metres across, with lifespans long enough to be meaningfully observed. These are precisely the conditions under which dragon sightings were historically clustered.

As discussed in earlier chapters on atmospheric plasma ecology, such conditions represent regions of enhanced energetic stability for plasma-based life. Ancient Chinese chronicles for example, describe dragons emerging from clouds, coiling around pillars of rain, or descending during lightning storms.

In Japan, the Ryū were described as radiant serpents capable of summoning wind or fire. Among the Norse and early Anglo-Saxons, fiery serpents flying in the sky heralded major atmospheric upheavals. In each case, dragons are situated in precisely the energetic environments where plasma lifeforms would be most active.

Their ancient ecological niche matches the very conditions favourable to plasma entity manifestation. This alignment between plasma ecology and mythic geography is so consistent that it becomes difficult to attribute the parallels to imagination alone.

Cross-Cultural Continuity

No theory of cultural diffusion can adequately explain why radiant serpentine sky-beings appear simultaneously in:

- Early dynastic China.
- Iron Age Europe.
- Ancient Mesoamerica.
- Aboriginal Australia.
- Ancient Egypt.
- Early India and Persia.

These civilisations lacked the geographical and linguistic pathways necessary for a single mythic archetype to spread globally, yet all describe the same morphology, the same behaviours, the same luminous properties, and the same auditory accompaniment of roars, thunder, humming, deep vibrations, or trumpet-like calls.

The convergence suggests a common environmental catalyst; the periodic appearance of serpentine plasma forms visible across the human perceptual spectrum. This same cross-cultural consistency appears in modern UAP reports, where elongated or serpentine luminous forms are still recorded, further supporting the idea of a stable underlying phenomenon rather than a human invention.

My own data strengthens this interpretation. Modern sightings of elongated plasma entities, some writhing, some columnar, others coiled like helices, mirror ancient accounts with uncanny fidelity. These forms frequently exhibit behaviours that appear intentional or responsive, a trait also ascribed to mythic dragons, which often act as guardians, teachers, adversaries, or omens.

Within the PEH framework, such responsiveness is not supernatural but ethological, reflecting the behavioural repertoire of a plasma organism interacting with human presence or atmospheric conditions.

Encounters Through Altered Perception

Across shamanic, mystical, and contemplative traditions, serpentine luminous beings do not appear only in the external sky. They arise just as vividly within visions, dreams, trance states, and altered modes of consciousness.

From the Amazonian ayahuasca cosmologies, where radiant serpents woven from fractal geometry are described as teachers or healers, to Tibetan visionary literature in which *"Wisdom dragons"* manifest as luminous, coiling currents of energy, the phenomenology is remarkably consistent.

These beings are radiant, fluid, serpentine, composed of interlocking patterns of light, and perceived as intelligent or intentional. Their descriptions correspond closely to the morphology of plasma entities as understood within the PEH. Plasma filaments naturally form braided, helical, and serpentine structures, often appearing fractal or geometric when viewed through shifting perceptual thresholds.

What is striking is that many visionary accounts describe patterns, lattices, coils, grids, and toroidal flows, instead of biological serpent bodies, forms that mirror the self-organising geometries observed in laboratory dusty plasma and in high-altitude transient luminous events.

Such convergence suggests that altered states do not generate fantasy imagery, but momentarily grant access to aspects of the plasma biosphere normally screened out by the brain's perceptual gating mechanisms. Human consciousness evolved to stabilise a coherent interpretive world from an overwhelming flux of sensory and energetic input.

Under certain conditions, ritual trance, entheogens, liminality, sensory monotony, fatigue, or meditative absorption, this gating loosens, and perception may widen to include electromagnetic structures ordinarily ignored or suppressed. In this context, the altered state inner vision, of a luminous serpent becomes less a hallucination and more an encounter with a plasma form whose presence is subtle, dynamic, and partially interdimensional.

As discussed earlier, the Ruliad provides a theoretical scaffold for such phenomena. If plasma entities inhabit computationally adjacent slices of reality, structures that coexist with ours but express themselves through different rules, frequencies, or update pathways, then the human mind may occasionally align with those slices.

Altered states create temporary resonances between our perceptual framework and the plasma entity's mode of existence, allowing brief cognitive overlap. This interpretation reframes visionary dragons not as metaphors but as perceptual encounters with real plasma intelligences, filtered through the symbolic architecture of the mind.

Just as certain animal forms appear more prominently during hypnagogic vision because they map well onto the neural structures involved in perceptual binding, serpentine plasma entities may appear in altered states because their underlying electromagnetic patterns interface cleanly with the brain's predictive machinery.

Hypnagogic and Hypnopompic States

Hypnagogic refers to the liminal state of consciousness and associated hallucinatory experiences, when falling asleep from an awake state. Hypnopompic refers to the similar state and associated hallucinatory experiences, when awakening. Both states are transitional, with vivid dream-like perceptions, sensations and sounds.

These are precisely the states in which neural filtering relaxes, and subtle electromagnetic cues become perceptible. Thus, the ancient and cross-cultural perception of luminous serpents of inner and outer space may represent two aspects of the same phenomenon.

When seen externally, these beings manifest as atmospheric plasma forms; when perceived internally, they appear as visionary serpents woven from light and geometry. Both may be valid expressions of contact with plasma-based intelligences operating at the boundary between the physical and the perceptual worlds.

This dual mode of appearance, external and internal, suggests that dragon encounters may occur across both atmospheric and perceptual domains, depending on the degree of alignment between environmental electromagnetic activity and human neural gating.

Such duality is entirely consistent with plasma structures that exist partly within the visible spectrum and partly within electromagnetic bands to which human perception is only intermittently sensitive.

Dragons as Memory of Plasma Ecology

When cultural ornamentation is removed, the core dragon motif stabilises into a remarkably consistent template; a radiant, serpentine, intelligent atmospheric being that interacts with storms, emits light, produces sound, and displays behaviours suggestive of agency.

Such traits are incompatible with any zoological organism but correspond naturally to high-energy plasma lifeforms. Many mythic dragons also guard treasure, a motif often interpreted symbolically. Yet within the PEH, treasure may refer to energetic nodes, resonance sites, or locations of electromagnetic significance.

The glowing orb so often held by Chinese dragons may symbolise the stable core of the plasma entity, akin to the luminous orbs frequently observed in filmed footage, nodes of coherence within an otherwise fluid plasma structure. Fire-breathing becomes plasma emission; flight becomes electromagnetic locomotion; serpentine form becomes field-constrained filamentarity.

Taken together, the global distribution, consistent morphology, environmental context, and recurring behavioural traits of dragons, all point toward a stable class of plasma lifeforms; intelligent beings that interacted with humanity across early and prehistory.

Rather than imaginative fantasy, dragon lore may represent humanity's earliest categorisation of the hidden plasma ecology surrounding it.

Conclusion

The dragon, far from a monstrous fantasy, may be humanity's earliest record of encounters with plasma-based life. These beings would have appeared at the thresholds of storms and auroras, in luminous coils across the sky, or as radiant serpents undulating at the periphery of human perception.

Ancient observers, lacking modern physics, translated these encounters into the lexicon of myth, yet the consistency of their descriptions strongly suggests a real, recurring phenomenon.

Today, with multi-spectrum imaging, high-speed sensors, and direct recordings of elongated, luminous entities, we are beginning to perceive what earlier cultures saw only briefly and interpreted through mythic forms.

The dragon becomes not a creature banished to the realm of legend but a plasma organism interacting with the Earth's atmosphere, an inhabitant of the hidden biosphere, whose presence was common enough to be woven into the fabric of global myth.

If plasma lifeforms exist, and the evidence accumulating across this book indicates they indeed do, then dragons were not imaginary beasts, but the first names ancient humanity gave to the serpents of light that shared the sky with them.

Their re-emergence in modern UAP sightings, atmospheric plasma events, and my own filmed encounters suggests that these beings never vanished at all. They merely remained unseen, awaiting a scientific framework capable of recognising them.

In this light, the dragon becomes the mythological façade of a real atmospheric organism, one whose physics we are only now beginning to understand. The myths preserved the form, and modern instruments are recovering the function. Together they reveal a plasma-based intelligence that has accompanied humanity from its earliest stories to its newest scientific inquiries.

21 Disruption, Humanity, and Contact

Until now, this book has focused primarily on observation; what anomalous luminous phenomena look like, how they behave, and how they may be understood within a plasma-based framework.

In several chapters, we have explored the possibility that certain unidentified anomalous phenomena (UAP) represent coherent, structured, plasma-based lifeforms rather than engineered craft.

Such a hypothesis carries implications that extend well beyond physics or classification. It raises an unavoidable question: what happens when observation gives way to interference?

Most historical encounters between humans and unidentified anomalous phenomena (UAP) have been passive. Witnesses report sightings, radar tracks, sensor anomalies, or fleeting visual impressions.

Even close-proximity encounters are typically described as observational rather than interactive. However, there are rare instances in which deliberate human action appears to have physically affected an anomalous object.

These moments are significant, not because they provide definitive answers, but because they expose how such phenomena respond to disruption.

THE YEMEN ORB VIDEO

One such case, briefly discussed in chapter one, emerged during a United States congressional oversight hearing in September 2025, when video footage captured by an MQ-9 Reaper drone on 30 October 2024 was presented.

The footage, often referred to as the "Yemen orb" or "Yemen footage", showed a high-speed aerial object being targeted and subsequently struck by a United States military AGM-114 "Hellfire" missile.

The object did not display the characteristics of a conventional missile or drone, nor did it behave in a manner consistent with known aerial vehicles following impact. Although the missile appeared to strike its target and cause partial fragmentation, the object continued on its trajectory with no obvious catastrophic loss of structural coherence.

Portions of the object were observed separating from the main body, yet its overall motion and persistence appeared largely unaffected. The publicly available footage does not allow for definitive conclusions regarding the object's composition, intent, or origin.

Nonetheless, the apparent survival of the phenomenon following kinetic impact invites careful consideration. If the object were a conventional craft, whether terrestrial or otherwise, such an impact would be expected to result in immediate incapacitation or destruction.

The observed persistence therefore suggests either an extraordinarily resilient engineered system, a misinterpretation of the footage, or a fundamentally different class of phenomenon.

Within the speculative framework developed in this book, one additional possibility emerges: that the object represents a plasma-derived, field-stabilised structure capable of maintaining coherence despite partial disruption. This leads to a further, more speculative implication.

If the structure observed is not a monolithic object but a distributed, self-organising system, fragmentation may not represent damage in the conventional sense. Instead, it may resemble division.

In biological systems, certain organisms exhibit non-localised organisation, where functional continuity is preserved across the whole rather than confined to a single central structure.

The hydra provides a familiar example; a simple organism capable of regenerating a complete body from a small fragment, because the genetic and organisational code is present throughout its form. The analogy is imperfect, but instructive.

If a plasma-based lifeform were organised through distributed electromagnetic coherence rather than through a centralised physical core, then fragmentation might not equate to destruction.

Under such conditions, separated regions could, in principle, re-establish coherence independently, leading to persistence or even replication. This concept aligns with earlier discussions of fractal recursion, nested resonant structures, and non-localised consciousness, where information and function are distributed across scales rather than localised in a single node.

It must be emphasised that this interpretation remains speculative. The available footage is limited, the context incomplete, and alternative explanations remain plausible. Sensor artefacts, misinterpretation of scale, or classified technological systems cannot be excluded.

The purpose of this discussion is not to assert a definitive conclusion, but to explore the implications if the plasma-entity hypothesis were correct. Beyond the physical response of the object itself, the incident raises a deeper issue.

The waters off Yemen, at the time of the recording, were considered an active operational environment. Military forces were operating under rules of engagement designed to protect commercial shipping and naval assets from missile and drone threats.

Within such a context, unidentified airborne objects may be treated as potential threats by default, particularly if they cannot be readily classified or ignored.

From a military perspective, such actions may be understandable. From an ontological perspective, however, they are unsettling. If unidentified phenomena are treated as hostile devices in the absence of clear intent, and if those phenomena are later found to represent a form of non-human intelligence rather than machinery, then the ethical stakes shift dramatically.

An act intended as neutralisation may instead constitute an act of harm toward an unfamiliar lifeform. This possibility reframes the broader discussion surrounding UAPs. The question is no longer limited to classification or threat assessment. It becomes a question of contact ethics.

At what point does defensive action become interference? At what point does interference become violence? How should uncertainty be managed when the nature of the phenomenon itself remains unresolved? The Yemen footage does not answer these questions. It merely exposes them.

It represents one of the few publicly acknowledged instances in which humans appear to have physically interacted with an anomalous aerial phenomenon in a deliberate and forceful manner. Whether the object was a device, a natural plasma structure, or something else entirely, the incident illustrates the limits of current interpretive frameworks.

If the phenomenon represents an intelligent, plasma-based lifeform, then attempts at destruction may be not only ineffective but counterproductive. Fragmentation might lead to persistence rather than elimination, and aggression might facilitate propagation rather than containment.

Even if this interpretation proves incorrect, the incident highlights a fundamental asymmetry; humanity possesses overwhelming destructive capability, but limited understanding of what it may be acting upon.

Human Interaction with Plasma-Like Phenomena

The Yemen incident is exceptional because it appears to involve deliberate kinetic interference with an unidentified aerial phenomenon. However, it is not the only context in which human activity appears to have physically influenced anomalous luminous phenomena.

When examined collectively, a number of well-documented cases suggest that certain atmospheric or plasma-like phenomena respond measurably to electromagnetic probing, energetic proximity, or partial disruption, even in the absence of overt destruction. These cases do not establish intent, intelligence, or life.

What they do establish is reactivity, behaviour that changes in response to human-generated fields or energy inputs. Within the framework explored in this book, such responsiveness is a necessary precondition for any discussion of interaction, whether biological, physical, or informational.

Radar and Electromagnetic Interaction

Among the most studied examples of non-destructive interaction is the long-standing phenomenon observed in the Hessdalen Valley, where luminous aerial lights have been reported continuously since the early 1980s.

Unlike many transient UAP cases, the Hessdalen lights have been subjected to repeated, instrumented scientific investigation over several decades. Researchers have documented that the lights frequently respond to active radar illumination.

Changes in velocity, brightness, altitude, and trajectory have been observed coincident with radar tracking. In some cases, the lights appear to split into multiple luminous bodies or recombine into a single structure.

Instrumentation deployed in the valley has recorded radio-frequency emissions, ionisation effects, and plasma-like optical spectra. The geological context of Hessdalen is notable. As discussed in an earlier chapter, the valley contains mineral-rich substrates, including copper, sulphur, and iron, along with conductive soils and persistent atmospheric inversions.

These conditions are capable of sustaining long-lived electrical gradients and plasma-supporting environments. The persistence of the phenomenon over decades suggests not a singular event, but a stable environmental system capable of repeatedly generating structured luminous forms. From the perspective of the plasma-entity hypothesis (PEH), Hessdalen represents a critical analogue.

Active electromagnetic probing alters the behaviour of the phenomenon without destroying it. This implies field sensitivity, coherence under perturbation, and a capacity to respond dynamically to external electromagnetic inputs. Whether this response is purely physical or represents a more complex organisational process remains unresolved, but the interaction itself is undeniable.

Weapon Proximity and Near-Miss Encounters

Historical military aviation provides another class of interaction: proximity to high-energy systems without confirmed impact. During the Second World War and the

Vietnam conflict, pilots repeatedly reported luminous objects pacing aircraft, hovering near engines, or manoeuvring alongside wings.

These phenomena, often referred to colloquially as foo fighters, were described as responsive to aircraft movement yet resistant to conventional engagement. In multiple accounts, pilots reported firing upon these objects or flying directly through their vicinity.

No debris, explosions, or radar-confirmed impacts were recorded. Instead, the objects either vanished abruptly, altered configuration, or repositioned beyond engagement range. The lack of destructive outcome suggests that either the objects were not materially vulnerable in a conventional sense, or that the interaction did not involve direct physical contact.

Within a plasma-based framework, these encounters may represent field disruption rather than structural damage. High-energy exhaust, electromagnetic emissions from engines, and weapons discharge could perturb a plasma structure sufficiently to alter its coherence or visibility without destroying it. The observed indifference to near-misses and apparent non-ballistic evasion align with this interpretation.

Atmospheric Electrical Interaction

Ball lightning remains a controversial and poorly understood atmospheric phenomenon, yet it provides valuable physical analogues for plasma coherence under disruption. In a number of documented cases, luminous spherical phenomena have been reported passing through solid materials, deforming metal, or leaving burn patterns inconsistent with conventional thermal damage.

In rare instances, ball lightning has been observed to fragment into multiple smaller luminous bodies before recombining or dissipating. These reports demonstrate that plasma-like structures can survive partial disruption, maintain coherence without rigid boundaries, and interact with matter in non-Newtonian ways.

While ball lightning is not necessarily proposed here as a lifeform, its behaviour establishes that physics permits stable, self-organising plasma structures capable of persistence, division, and recombination under certain conditions.

Such behaviour closely parallels the fragmentation-without-destruction pattern implied in the Yemen footage. It reinforces the plausibility that partial disruption of a plasma-derived structure does not necessarily equate to termination.

Directed Energy and Laser Illumination Responses

Scattered civilian and military reports describe anomalous aerial lights responding to laser illumination. In these cases, coherent light input is associated with changes in brightness, abrupt motion, or sudden disappearance.

Notably, these effects are not accompanied by observable thermal damage, suggesting that the interaction is not driven by heating. Instead, such responses may indicate sensitivity to coherent electromagnetic input, phase disruption, or resonance interference.

In plasma systems, even modest electromagnetic perturbations can alter stability, confinement, or emission characteristics. The consistency of reported responses across unrelated incidents suggests that these phenomena, whatever their ultimate nature, exist within an electromagnetic regime where external fields matter.

Fragmentation Without Destruction

The "Yemen orb" footage occupies a unique position among interaction cases. It may represent the only publicly acknowledged instance in which a kinetic weapon appears to have impacted an unidentified aerial phenomenon, resulting in partial fragmentation without obvious loss of functional coherence.

This places it in a rare category shared only loosely with ball lightning fragmentation events and occasional Hessdalen splitting observations. The significance of this pattern lies not in its dramatic nature, but in its implications. Fragmentation without destruction challenges object-based assumptions of integrity and failure.

It suggests that the phenomenon involved may be organised through distributed coherence rather than through a centralised physical structure. Within the speculative framework developed earlier in this book, such behaviour is consistent with non-localised organisation and fractal recursion.

If coherence is maintained across nested scales, then disruption of one region does not necessarily compromise the whole. In extreme cases, disruption may even enable persistence through division.

Framing the Implications

Taken together, these cases do not prove the existence of plasma-based life. They do, however, establish that human activity can physically influence plasma-like anomalous phenomena without destruction and that such influence may produce adaptive or realignment responses rather than catastrophic failure.

This places humanity in a precarious interpretive position: capable of intervention, but uncertain of what it is intervening upon. We possess the means to disrupt phenomena we do not yet understand, but lack the conceptual frameworks to determine whether such disruption constitutes interference, harm, or communication.

If some subset of these phenomena represent organised, persistent systems, whether biological, proto-biological, or something else entirely, then the transition from observation to interaction carries ethical and scientific consequences that have yet to be fully considered.

The Yemen incident may therefore be less important as an act of force than as a boundary marker; a moment where a lack of understanding met intervention. In that sense, it belongs not only to discussions of propulsion or threat, but to a deeper inquiry into how humanity engages with phenomena that occupy the margins between physics, environment, and possible life.

As encounters with unidentified phenomena continue, the distinction between observation and interference will become increasingly important. The transition from watching, to acting, carries consequences that extend beyond defence or security. It touches upon humanity's readiness to engage with forms of intelligence that do not conform to biological, mechanical, or cultural expectations.

In the chapters that follow, we will return to the mechanisms by which plasma-based entities might interact with physical environments, human perception, and technological systems.

The Yemen orb incident serves as a boundary marker within that exploration; a moment where ambiguity and force intersect. Whether it represents fear of the unknown, misinterpretation, or misclassification, it underscores the need for caution, restraint, and deeper inquiry before action replaces understanding.

22 Crossing the Boundaries

If plasma lifeforms exist as coherent, intelligent structures of plasma and electromagnetic energy, as this book proposes, then certain questions arise with increasing urgency. If they exist, what of their proximity, their purpose, and their influence? Have they always been here? Have they affected us from behind the veil of perception?

Or, more provocatively, have they crossed the boundary into our realm in ways far more intimate than fleeting lights in the sky? This chapter explores a radical possibility. Plasma-based lifeforms may not be limited to distant atmospheric displays or ambiguous photometric signatures.

Instead, they may be capable of engaging with the human world in deeply personal, psychological, and occasionally physical ways. These interactions span a spectrum ranging from the subtle and cognitive to the intrusive and visceral. If these beings are real, and if their mode of existence depends upon electromagnetic architecture rather than biological tissue, then the ways in which they cross into human experience may defy our conventional categories of contact.

Most encounters reported throughout history involve lights in the sky, strange luminous forms, unexplained atmospheric disturbances, or interactions that distort electromagnetic equipment. Yet a persistent subset of accounts describe entities entering more intimate proximity: moving into our homes, our bedrooms, our perceptual field, our emotions, and at the most contentious boundary, our bodies and minds.

These reports are often dismissed as hallucinations, sleep paralysis, or psychological stress. In many instances such explanations may be appropriate. However, when examined through the lens of the plasma entity hypothesis (PEH) and the idea of a non-visible energetic biosphere, these experiences may take on a new coherence.

If plasma entities interact through field entanglement, in the sense of classical electromagnetic coupling and resonance, or subtle entrainment, then physical contact does not require biological flesh, nor does cognitive influence require supernatural models.

What humans have interpreted as spirits, guides, guardian angels, demons, or ancestral forces may instead represent the interface between a non-biological plasma intelligence and the electrochemical consciousness of the human brain.

This chapter examines several modes of interaction: physical impressions, energetic imprints, telepathic or quasi-telepathic communication, cognitive resonance, emotional influence, and the gentle or forceful steering of behaviour. These cases challenge the assumption that intelligence must be material to be real, and that influence must be visible to be profound.

If plasma entities operate through nested electromagnetic fields, as described in earlier chapters, then interacting with human neurophysiology may require nothing more than frequency alignment, the right environmental conditions, and a moment of resonance. In such circumstances, the boundary between our world and theirs becomes porous.

We begin at the most visceral edge of this spectrum:

Physical Interaction and the Doris Bither Case

One of the most controversial and distressing examples of alleged physical contact is the case of Doris Bither, often sensationalised as "*The Entity Case*". Investigated in the 1970s by parapsychologists Kerry Gaynor and Barry Taff, the case featured violent poltergeist activity, electrical disturbances, and repeated sexual assaults by unseen forces.

Bither reported that these entities were invisible yet strong enough to leave bruises, hurl objects, and manipulate her environment. During episodes of heightened activity, the investigators photographed luminous forms appearing above Bither, hazy and orb-like, often structured enough to suggest localised concentrations of energy.

In the context of the plasma entity hypothesis (PEH), such manifestations may represent a plasma entity temporarily condensing into a denser state of ionisation or coherent charge, a phenomenon observed in laboratory plasmas when they undergo sudden transitions or reorganisations.

The physical effects described, cold zones, displaced objects, pressure on the body, and bruising, could be produced through directed electromagnetic force. If an entity can manipulate local charge density or generate short-lived plasma sheaths, then exerting mechanical pressure on matter becomes feasible.

This does not imply malign intent, but it does illustrate a form of contact that crosses the boundaries between atmospheric phenomena and human physical experience. Although the Bither case is extreme, it is not unique.

Similar accounts of bedroom visitations, paralysis accompanied by luminous presences, and sensations of being touched or pressed upon appear across cultures and epochs. These experiences edge into the territory of incubus and succubus traditions, which we will return to below.

Min Min Lights and the Behavioural Frontier

The Australian Min Min lights offer a culturally grounded, geographically consistent example of plasma-like entities interacting within human proximity. Aboriginal Australians have described them for millennia, usually as glowing orbs that follow travellers, hover above ground level, or approach before retreating.

They are often associated with a sense of unease, dread, or compulsion. Anthropologists recording Warlpiri, Kalkadoon, and Pitta Pitta accounts note that Min Min lights sometimes appear to *"call"* people, drawing them forward or leading them astray. Some witnesses describe the onset of altered perception, as if their attention has been captured or entrained.

These descriptions strongly parallel the psychological effects reported in modern UAP encounters. From a plasma perspective, the Min Min lights are particularly interesting because they occur in arid environments with high charge gradients between ground and sky, conditions ideal for stable low-altitude plasmas.

Their behaviour, such as hovering, pursuit, retreat, and apparent interest in observers, is consistent with an entity whose motion is governed by electromagnetic coupling rather than aerodynamics. In some cases, witnesses report physiological sensations: tingling, heat, pressure, or emotional shifts.

These could reflect direct interaction with electromagnetic fields, especially if the plasma entity is adjusting its configuration in response to human presence. However, not all Min Min interactions are negative. Some Aboriginal stories describe the lights as warning spirits or guardians, echoing cross-cultural accounts of guiding luminous beings.

This diversity mirrors the behavioural spectrum observed in plasma entities elsewhere: curiosity, approach, retreat, and sometimes a form of reciprocal attention.

Ancient Near Eastern Parallels

To understand the historical depth of this phenomenon, we must turn to the oldest demonological traditions, particularly those of Mesopotamia.

One of the most persistent and misunderstood figures in world folklore, *Lilith*, originates from the Akkadian word lilītu, which itself derives from *lilû*, meaning *"wind spirit", "airborne being"*, or *"night-dweller"*.

These entities were described as energetic beings inhabiting the liminal zone between the physical world and the atmospheric or spiritual domains. The lilītu were associated with storms, nocturnal visitation, erotic encounters, and the ability to influence or afflict sleepers.

They were described as luminous, ephemeral, and capable of appearing in the threshold states between waking and dreaming. In many early incantation texts, they are said to glide, drift, or *"move in the air like wind"*, which is striking when compared with descriptions of plasma forms.

Over centuries the figure was mythologised into Lilith, a night spirit associated with seduction, intrusive dreams, erotic compulsion, and what later Christian writers cast as demonic possession. Yet the earliest sources emphasise her atmospheric nature: not a demon of flesh, but a being of air, light, or wind, capable of influencing human minds and bodies through proximity.

Under the PEH framework, the lilītu traditions can be reinterpreted as early ethnographic observations of plasma entities interacting with humans during liminal states.

INCUBUS, SUCCUBUS AND THE PLASMA INTERFACE

The medieval incubus and succubus traditions represent a continuation of this ancient lineage. These beings are said to visit sleepers, induce paralysis, provoke erotic encounters, manipulate dreams, or exert psychological influence.

In modern clinical language, these events are often categorised as sleep paralysis hallucinations. Yet the consistency, antiquity, and cross-cultural spread of such narratives suggest a deeper root.

If plasma entities are capable of partial condensation, proximity-based resonance with neural oscillations, or influencing local bioelectric fields, then reports of nocturnal visitation take on a new significance. Several elements of incubus and succubus encounters parallel documented features of plasma entity interaction:

Sleep Paralysis and Electromagnetic Disruption

Many sleepers report feeling paralysed, pressed down, or unable to speak. Under sufficiently intense or localised field conditions, electromagnetic fields are known to disrupt motor cortex activity and induce transient paralysis.

Within the plasma entity framework, such effects need not be pathological. A nearby coherent plasma structure, particularly one undergoing partial condensation or field reconfiguration, could transiently interfere with motor signalling while leaving consciousness intact.

This would produce the characteristic terror of paralysis combined with heightened awareness reported in incubus encounters. The experience may therefore reflect a physiological response to external electromagnetic influence rather than a purely internal hallucination.

A Presence Sensed before Manifestation

Witnesses often describe the impression of an intelligence in the room before any form appears. This parallels reports from Yakima, Min Min interactions, and UAP witnesses who felt they were being called or watched.

From a plasma perspective, this suggests that field interaction precedes visual perception. Human awareness may respond first to subtle shifts in local electromagnetic conditions before the entity becomes optically visible. In this sense, the sensation of presence may be the earliest stage of boundary crossing, occurring at the level of cognition rather than sight.

Luminous or Shadowlike Forms

Some accounts describe beings appearing as glowing shapes, flickers of light, or amorphous shadows. Such variability is consistent with plasma structures operating near the threshold of visibility.

Minor changes in charge density, temperature, or environmental coupling could cause an entity to appear radiant, translucent, or shadow-like. The lack of stable morphology reported in incubus traditions mirrors the polymorphic behaviour also observed in modern plasma phenomena and UAP encounters.

Erotic Contact or Emotional Resonance

This is the most contentious element of the tradition, yet from a plasma perspective, such experiences could arise from entrainment of neural circuits governing arousal, emotion, and sensation. Witnesses often describe lingering anxiety, exhaustion, heightened sensitivity, or a sense of personal transformation.

These mirror after-effects of some UAP encounters documented in modern reports. Seen through the plasma entity hypothesis (PEH), the incubus and succubus phenomena are not demonic intrusions but early cultural interpretations of plasma entities interfacing with human consciousness and physiology.

The erotic framing may simply reflect the human brain's attempt to interpret unfamiliar sensory input, particularly when occurring during sleep-state boundary conditions.

Cognitive Influence and Thought Resonance

Perhaps the most widespread yet least recognised mode of interaction is subtle cognitive influence. This may appear as a fleeting thought, a sudden emotional shift, an irrational urge, or a sense of intuition that seems to come from outside ordinary mental processes.

If plasma entities possess structured electromagnetic fields capable of coupling with neural oscillations, then cognitive entrainment becomes a plausible mechanism for influence.

Such interactions would necessarily be highly constrained, as electromagnetic influence attenuates rapidly with distance and is strongly shaped by biological shielding, environmental noise, and competing field sources; meaningful coupling would therefore require close proximity, specific field configurations, or transient conditions of heightened sensitivity.

The human brain is highly sensitive to external fields, especially at low frequencies. A modest change in local charge distribution can alter attention, decision-making, or emotional tone. Most of these influences would feel internal, indistinguishable from one's own thoughts.

When positive, such experiences are often interpreted as insight or inspiration. When negative, they may be dismissed as intrusive thoughts or anxiety. The boundary between self and other becomes porous. This concept parallels ancient beliefs in guiding spirits, guardian angels, muses, daimons, or ancestors who whisper ideas or guidance.

It also resonates with Carl Jung's understanding of the collective unconscious and synchronicity. What appears to be inner dialogue may sometimes be interaction with a larger energetic field that includes plasma entities.

Modern Contact Narratives

Modern UAP encounters reflect the same behavioural palette found in ancient traditions. During the Yakima Valley flap of the 1970s, several witnesses reported a sense of being summoned outdoors moments before glowing orbs appeared. Similarly, Min Min witnesses often describe a compulsion to follow the lights.

The Phoenix Lights event of 1997 also included reports of intense emotional responses, fear, tranquillity and awe, that were disproportionate to the visual stimulus, suggesting a possible influence on limbic or cognitive systems. These accounts demonstrate that plasma entities, if that is indeed what these phenomena represent, may interact across a continuum:

- Atmospheric or luminous manifestation.
- Electromagnetic disturbance.
- Emotional resonance.
- Cognitive entrainment.
- Sleep-state infiltration.
- Physical influence.

This continuum mirrors the full range of incubus and succubus traditions, the lilītu of Mesopotamia, Aboriginal Min Min encounters, biblical angelic visitations, and many modern UAP reports. The phenomenon is polymorphic. Its manifestations span visual, audible, cognitive, emotional, and physical domains.

To force these experiences into rigid categories of physical or psychic is to miss their unity. Plasma entities are not crossing a single boundary; they are crossing every boundary, shaping human perceptions, myths, fears, insights, and decisions.

Towards a Broader Ontology

If plasma entities exist, and if they possess the capacity for influence across the electromagnetic substrate of human consciousness, then we must accept a broader ontology. Life may not be limited to carbon and water. Intelligence may not be confined to biological tissue.

Communication may not require language. Contact may occur in a dream, a thought, a flash of insight, an inexplicable urge, or a subtle emotional shift. Some interactions may be benevolent, others neutral, and others intrusive. This is consistent with all known ecosystems; diversity of behaviour emerges naturally across species.

The watchers of ancient texts, the radiant beings of religious visions, the serpents of shamanic trance, the incubus and succubus traditions, the Min Min lights, and the orbs of modern UAP sightings may all represent different cultural lenses directed at the same underlying plasma intelligence.

If so, then our history, mythology, and perhaps even aspects of our psychology have been shaped in part by beings operating across a parallel electromagnetic ecology that intersects intermittently with our own.

Not every thought may be our own. Not every decision wholly human. Plasma entities may be more than visitors. They may be neighbours, participants, or even subtle guides, influencing humanity through resonance rather than force, shaping our myths and perhaps even our inner lives.

In those rare moments when they cross fully into our physical realm, when they touch, speak, or occupy our space, we return to the fundamental question; do they act as intruders, or as members of a parallel layer of existence with whom we have always shared this world? Their message may be one of fear or one of reverence.

Their presence may evoke awe, dread, fascination, or transformation. Yet beneath all these reactions lies a deeper truth: contact between plasma entities and humans may be far more ancient, widespread, and intimate than we have ever dared to consider.

If this interpretation is correct, then what we call paranormal intrusion, religious revelation, shamanic communion or UAP encounter may not represent different categories of event, but different degrees of permeability between electromagnetic systems: ours, grounded in neural chemistry, and theirs, expressed through coherent plasma.

In this light, the boundary between human cognition and external intelligence becomes a gradient rather than a wall. Interaction emerges wherever frequencies meet, just as resonance bridges the gap between sound, structure and vibration. The question is not merely whether plasma entities cross into our world, but how often they already do so without our recognition.

Their presence forces us to reconsider the nature of perception itself, suggesting that the human mind may be less an isolated organ and more a participant in a wider energetic ecology, one whose inhabitants have been interacting with us since prehistory, shaping experience in ways both subtle and profound.

23 Storms as Sentient Structures

The sky was already ablaze when our flight finally touched down on the 13th of April 2020, long before I began this incredible journey of discovery. We had been delayed on the tarmac in Perth for reasons unstated and arrived to find the distant storm in full force, towering above the mine camp, its swollen mass reflecting the hue of the setting sun against a flickering orange horizon.

From the airport, some eight to ten kilometres away, I raised my mobile phone and began to record. I didn't know it at the time, but those 39 seconds would capture something extraordinary. The storm cloud was immense, layered and dense with electrical activity. It stretched across the sky like a vast, suspended neon sign.

Based on scale comparisons and the known size of the camp, I estimate the cloud to have been between three to four kilometres in height and eight to ten kilometres long. Yet despite the magnitude and intensity of its internal lightning, the storm mass held its form. Steadfast and motionless, like an immense battleship anchored in the sky.

For years afterwards, the footage remained nothing more than a wonder of nature, contained in a digital file. After beginning this research, extreme weather events became more than just a casual interest. I revisited many such videos with a much more critical eye, and I returned to that video with particular curiosity.

In isolated frames, revealed only during certain internal flashes, there appeared a huge, sharply bounded cuboid embedded within the cloud, the shadow it cast within the storm body I estimate to be one hundred and fifty to two hundred metres across, with geometry that should not exist within vapour.

Over the course of the recording, more than thirty distinct internal discharges occurred within the cloud. Each flash revealed stark contours, chambers, huge beams of shadow and repeated coherent shapes that suggested structure rather than chaos. During this intense electrical activity, the cloud did not drift or distort.

As I studied more frames, the impression became unmistakable, this was not a weather system merely reacting to thermals or barometric shifts. This storm cloud appeared to have structure, massive and deliberate, maintaining shape and position with improbable coherence while brilliant flashes of internal lightning rippled beneath its vaporous skin.

The storm cloud was not disrupted by the heat of so many massive discharges. Moreover, it seemed alive in some sense of the word. At a glance, it appeared to be a high-altitude storm cell, but over time it exhibited behaviour that storm systems should not.

Internally lit by a near-continuous series of electrical discharges, the entire volume of the cloud would pulse like a living organism, shadows flickering within like chambers or organs reacting to waves of energy. These were not the jagged, atmospheric lashes of fork lightning striking out or downward, just contained, rhythmic releases of energy, as though some vast intelligence was metabolising the very atmosphere.

Around it, what seemed to be the same faint entities I had been investigating, moved away at speed, in the periphery. Had they been drawn to the great structure, fed and were now going about their business in a revitalised manner? Whether they were feeding, communicating, or emerging from it remains uncertain.

Under normal atmospheric conditions, such repeated lightning activity within a storm cell would destabilise the clouds structure. The thermal energy released, often exceeding 30,000 °C per strike should cause rapid expansion, vertical shearing and dissipation of water vapor. Yet this cloud retained its form.

It neither grew nor contracted in the expected ways. It hovered in place over the camp, unmoved by prevailing winds, as if held by a force greater than convection or pressure differentials. Science tells us that lightning arises from charge separation within storm clouds. Ice crystals, supercooled water droplets, and graupel exchange electrons as they rise and fall in turbulent updrafts.

Discharges occur when the electric field exceeds the dielectric breakdown of air, in a similar way to the coronal discharges observed within willy-willies. This is true, but such an explanation did not seem to account for what I had observed, the apparent conservation of form, the self-regulation of heat, the non-destructive internalization of electrical events.

A new possibility presents itself; what if this enormous storm cloud was not merely a host for the electrical discharges of the storm, but had also become host to an energy based intelligence? A vast, kilometres long plasma lifeform able to maintain the coherence of the cloud, remaining embedded and cloaked within until whatever business that it was conducting, was complete.

Theoretically, was it using its own electromagnetic fields to organize vapor and atmospheric charge into a stable body-form? If this was so, then this was not just a storm cloud, it was a Storm Body, a living structure sustained by controlled lightning, electromagnetic manipulation and possibly a form of bioelectric metabolism that spans the boundary between the organic and the non-material.

Either I was misunderstanding what the physics predicts at this scale, or an additional organising factor was present. To understand how such a being could endure, one must first confront the paradox of plasma itself; it is at once the most abundant state of matter in the universe and the most difficult to confine.

On Earth, we encounter plasma fleetingly in lightning bolts, in auroras, in the flicker of a neon light and very occasionally, as ball lightning. It is brilliant, energetic, unstable, and yet, within this thundercloud the plasma behaved like a lattice. It contained and distributed energy without disrupting its distinct cloud formation. In traditional atmospheric physics, lightning is a product of instability.

Ionised channels called stepped leaders are said to propagate in abrupt jumps through a field of oppositely charged air, breaking down the dielectric strength of the atmosphere. A return stroke, a flood of electrons from the ground or opposing charge centre, surges back along the path, heating the air to extraordinary temperatures.

This heat, in turn, causes explosive expansion within the air causing a sonic boom we call thunder and would normally lead to the rapid disassembly of any vaporous form in its path. On this occasion in the Pilbara, the opposite occurred. The more lightning was discharged within the cloud, the more stable the structure seemed to be, as though the entity was using energy not destructively, but as structural reinforcement.

Plasma physics offers a precedent for this behaviour in the form of the Z-pinch effect which may be controlled in laboratory conditions, but is uncontrolled in lightning. In laboratory conditions and fusion experiments, plasma can be contained and focused by its own electromagnetic fields.

When electric current flows through a plasma filament, it generates a circular magnetic field. This field compresses the plasma inward, pinching it into a coherent line. It is this very mechanism that keeps lightning bolts narrow and fast-moving, despite their violent birth.

Imagine if you will, a being of colossal size, kilometres in extent, capable of using the Z-pinch effect to stabilise its own internal structure, looping currents like muscles contracting. Therefore, what meteorologists interpret as turbulence, current, or stepped-leader propagation, might be something more than just chaotic discharge, instead, the intense discharges here acted with a definite function.

Were they maybe analogous to nervous system function, or structural reinforcement? Some theoretical and experimental laboratory research has shown that plasma confined within geometric cavities can, under certain conditions, enter resonant states of stability. In such configurations, polyhedral plasmas may store charge, regulate ion flow, and behave in ways analogous to information lattices, exhibiting coherent and sustained resonant behaviours.

If such a lattice were to form, whether through self-organisation or deliberate regulation, it could explain the unusual endurance and structural stability of the Pilbara storm body. The cuboid may have served as a capacitor, a gestation node, a structural anchor, or even a resonant chamber for energy processing, but this observation, compelling as it is, is only one piece of the puzzle.

When I later viewed other videos of storms I had filmed, once again I observed something I had not expected; entities moving at speed, seemingly away from major electrical discharges. They did not drift randomly like debris caught in wind, they moved with intention, at speeds in the thousands of kilometres per hour range, as if fully charged. This velocity was estimated using the framerate of the video and the estimated distance travelled.

In one video, after closely viewing it many times and observing the object travelling at high speed, south to north, I subsequently found more frames where the object had travelled in a huge loop, doubling back over the field of view, then off in a west to east trajectory, possibly in the direction of the next major lightning strike.

I had missed this second pass, even after numerous viewings over several months, closely inspecting many individual frames. This is a good example of how challenging these lifeforms can be to find, but also raised another possibility. Maybe they are able to sense the increasing charge and anticipate the position of an electrical discharge before it occurs.

In a biological ecology, apex predators discern feeding cycles and patterns. We know orca specifically gather to target seals by "wave-washing". They anticipate humpback whale migration routes, waiting to separate calves from the herd. Similarly, lions know when to target prey at watering holes.

These behaviours are not random, they are learned predatorial behaviours. I believe I filmed what may be a parallel behaviour in electrical storms. Plasma entities speeding toward and from gradients of charge, assembling where the coming lightning would deliver harvestable energy. Perhaps lightning is feeding time, which brings us to our next subject.

Ball Lightning

This exceptionally rare and elusive phenomenon is often seen as a glowing orb of varying colours and has been reported throughout history. Sometimes reported to pass through windows and walls, or as floating lethargically, occasionally exploding with tremendous violence.

Ball lightning has also been reported on an aircraft, en route from New York to Washington in 1953 by R.C. Jennison. After the aircraft was enveloped by a huge electrical discharge, a brightly glowing orb, approximately twenty centimetres in diameter, entered the passenger compartment from the captain's cabin and slowly travelled the full length of the aircraft.

We have discovered in previous chapters, how drones, dragons, gremlins, foo fighters, UFOs and UAPs, may have been misnomers for plasma lifeforms. Even some videos which show what are purported to be ball lightning high in the atmosphere, are instantly recognisable to me as plasma entities, though

I don't doubt the existence of ball lightning; indeed, I suspect it is connected to the plasma entity hypothesis (PEH). Science currently struggles to explain how a luminous plasma sphere could move slowly, remain coherent, and even penetrate solid objects without breaking them.

However, what if some instances of ball lightning are not merely transient phenomena, but represent a particular energetic state, an over-charged plasma organism engorged with energy, struggling under metabolic load, then its behaviour makes biological sense, just like a predator after feasting.

This interpretation remains speculative, but it accounts for the motion of the orb often seen as lethargic and the coherence unstable. Some may successfully dissipate their excess charge and return to their mostly invisible state, while others explode, unable to control the excess.

This may be due to environmental effects, location, temperature and humidity, atmospheric pressure or any number of other variables. Some reportedly, pass through walls and windows.

The Illusion of Solid Matter

Ordinary matter is not solid; this is a well-established physical principle within the scientific community. An illusion caused by electromagnetic repulsion between electron shells. Ball lightning as well as orbs and the many legends and myths of ghosts reportedly walking through walls and solid objects may be a scientifically sound ability.

To pass through matter, an object does not need to become ethereal, it only needs to neutralise its surface field. If a plasma lifeform can modulate charge, invert polarity, reduce surface repulsion, synchronise phase with the electron density of a barrier, then in theory, it could slide through, just like in all those ghost stories, but supported by real physics.

In its baseline state, lean, low-charge, minimal luminosity, such a being would move quickly, effortlessly, almost invisibly. In its overcharged state, bloated, bright, heat-laden, it would move slowly, pass through solids unpredictably, and possibly risk catastrophic failure. So, perhaps ball lightning is simply the plasma equivalent of over-eating.

This raises a deeper question; are some storms opportunistic habitats these lifeforms exploit, or are storms created within their leviathan, ethereal bodies? Biology offers precedent, where beavers do not wait for rivers to form natural pools, they make their own dams. Termites don't look for a home at the optimal temperature; they build air-conditioned towers. Coral polyps do the same, constructing vast reefs that can be seen from space. Life shapes its environment into a body.

If plasma beings are atmospheric intelligences, perhaps they do not merely feed within storms, perhaps they construct storms. To do so, an entity would need only three abilities; to control charge distribution, ionise air to create condensation nuclei, and induce thermal uplift. Pre-industrial cultures, often regarded as primitive by modern standards, and whose shamans have performed rain-summoning rituals for millennia, may well have been interacting with something real.

If this theory is correct and such beings exist, they may operate in two modes: as opportunists, emerging when atmospheric parameters align, or as architects, actively generating storms to meet their energetic requirements. In this model, a translucent plasma entity could remain largely dormant and invisible, awaiting the convergence of humidity, charge gradients, and magnetic tension.

When these conditions align, it emerges with intent. Viewed in this light, mythology is reframed. Gods who wield lightning, dragons that dwell within clouds, and sky-serpents whose movements foretold rain may not be metaphors, but observations filtered through limited vocabulary and cultural context.

If plasma can form coherent structures, store energy, distribute heat, regulate charge, and respond dynamically to its environment, then by strict definition it satisfies more criteria for life than some entities we already classify as living, including viruses. This chapter began with a camera held by someone who believed he was filming a storm.

Storms, at least some storms, may not merely be weather systems, they may be non-biological organisms. Immense, radiant, and sentient in ways alien to known biology. Perhaps when lightning flashes in the night sky, we are not witnessing imbalance correcting itself, but pulse, respiration, metabolisation.

Perhaps when thunder rolls, we are hearing the pressure-signature of life. From the perspective of modern science, such beings may seem improbable, even impossible but the phenomena are real. If the phenomena are real, it can be confirmed by observation, measurement and analysis.

Historically, science has always evolved by confronting what was once considered implausible or impossible. From Robert Bakers theory that birds were linked to dinosaurs and J. Harlen Bretz's theory of catastrophic floods shaping geological formations, initially rejected for 56 years, to Alfred Wegener's theory of continental drift which was dismissed for lack of a plausible mechanism for continental movement, to my personal favourite, the Kordylewski clouds first reported in 1961 and confirmed 57 years later in 2018.

It has been mentioned in many other books, but for those few who haven't heard it, I will include it here: "There are three stages of acceptance. The first is ridicule or denial; then comes acknowledgement as true but not important; finally, there is acceptance as both true and significant."

Perhaps we are not alone in the way we thought. Perhaps the skies are not empty, but full and inhabited by forms of conscious awareness. Not invaders, nor visitors, instead, ancient dwellers of the electromagnetic substrate, belonging here and emerging from an invisible, hidden biosphere. A realm only hinted at, rarely glimpsed by man.

When we speak of weather, perhaps we are speaking unknowingly, of a language spoken by these entities. The thunder is not just pressure release. The lightning is not just imbalance. These may be signals, pulses, or lifeblood of beings whose nature we are only just beginning to uncover. When we look up and see a storm, perhaps just once, we should look up and wonder if the storm is looking back.

Now, with hindsight shaped by study, field experience, and a growing body of anomalous evidence, I see this subject differently. With this insight, I regard that immense cloud as an organism of leviathan proportions and possibly an interface through which humanity is monitored, communicated with and influenced, not a random confluence of pressure gradients and solar heating.

24 Conclusion

Toward an Understanding of the Hidden Biosphere

This book began with a simple question:

"What if Earth harbours a parallel biosphere, largely invisible because it is electromagnetic rather than biological, not because it is alien?"

Over the chapters that followed, that question has been examined from multiple angles: observational, physical, theoretical, historical, and ethical. No single strand has been presented as decisive, instead, a pattern has been allowed to emerge. As that pattern has taken shape, the original question has naturally sharpened.

If such a biosphere exists, then many of the luminous anomalies observed in our skies, oceans, and atmosphere may not be machines at all, but manifestations of a non-biological ecology structured by fields rather than matter.

The central argument advanced here is not that all unidentified anomalous phenomena (UAP) represent plasma-based life, nor that alternative explanations such as advanced technology, misidentification, or natural atmospheric effects can be discarded.

Rather, it is that a specific subset of recurring phenomena exhibits characteristics that are poorly explained by existing frameworks, yet remain physically admissible within known, though underexplored, domains of plasma physics, electromagnetism, and complex systems theory.

Throughout this work, the emphasis has been placed on restraint. Extraordinary claims require extraordinary evidence, but they also require intellectual openness. Historically, many scientific advances have emerged from the willingness to question entrenched assumptions and to explore anomalous data without premature dismissal, not from the immediate availability of proof.

The hypothesis explored here occupies that intermediate space, speculative and unconventional but grounded within scientific boundaries.

From Objects to Systems

One of the most persistent obstacles in the study of UAPs has been the tendency to interpret all anomalous phenomena as physical objects. Craft are expected to have structure, propulsion, mass, and failure modes analogous to familiar machines.

When phenomena fail to conform to these expectations, they are often either forced into the extraterrestrial vehicle model or dismissed outright. This book has proposed an alternative framing. Many observed anomalies may be better understood as systems rather than objects; as field-stabilised, distributed structures whose coherence arises from electromagnetic interaction rather than mechanical assembly.

In such systems, it is not mass that governs behaviour, but resonance, charge distribution, and environmental coupling. Acceleration without inertia, trans-medium movement, low observability, and resistance to conventional disruption are not paradoxes in this context; they are expected properties.

By shifting the focus from craft to coherence, a number of long-standing puzzles lose their mystery. The absence of sonic booms, the lack of propulsion signatures, the ability to appear and disappear, and the tendency to fragment or reconfigure without destruction all follow naturally from plasma-based organisation rather than from exotic engineering.

Intelligence Without Biology

Perhaps the most challenging implication of this framework concerns intelligence itself. Human cognition is deeply shaped by biological precedent. We expect intelligence to be localised, cellular, metabolically driven, and bound by neural architecture.

Yet intelligence, at its most fundamental level, is not a substance but a process; the ability to store information, respond to stimuli, adapt to changing conditions, and maintain internal coherence. In this context, the possibility of non-biological intelligence becomes less radical.

A plasma-based system organised through nested resonant structures, fractal recursion, and distributed electromagnetic coherence could, in principle, meet these criteria without resembling life as we know it.

Such intelligence would not think in language, images, or linear sequences, but in harmonics, interference patterns, and field dynamics. Its perception would be immediate rather than mediated, its cognition parallel rather than sequential.

If such entities exist, they would not be visitors in the conventional sense. They would be *native to the environment that sustains them*; emerging where electromagnetic gradients, mineral substrates, atmospheric conditions, and energetic flux permit long-term coherence.

In this view, Earth is not merely a passive stage for biological evolution, but a complex electromagnetic ecosystem capable of supporting multiple forms of organised complexity simultaneously.

Observation, Interference, and Responsibility

As explored in the later chapters, the distinction between observation and interaction is not trivial. Human activity increasingly alters the electromagnetic environment through radar, directed energy, aviation, and weapon systems.

The evidence reviewed here suggests that some anomalous phenomena respond measurably to such perturbations. Whether these responses are purely physical or reflect a more organised system remains unresolved, but the implication is clear; interaction is already occurring, often unintentionally.

The Yemen orb incident serves as a stark example. Regardless of its ultimate explanation, it represents a moment where uncertainty became forcefulness.

If future investigation were to reveal that some anomalous phenomena constitute organised, persistent systems rather than inert devices, then the ethical framework surrounding engagement would require urgent revision.

The question would no longer be how we neutralise an unknown, but how do we avoid harm in the absence of understanding. This work does not argue for passivity, nor does it suggest that all unidentified phenomena are benign. It does, however, argue for epistemic humility.

Acting decisively in ignorance may offer short-term security, but it risks long-term misunderstanding. History offers many examples where unfamiliar life or complex systems were treated as threats simply because they were not recognised as such.

Why This Matters Now

The timing of this inquiry is not accidental. Advances in imaging technology have expanded human perception beyond the limits of unaided senses. Phenomena once invisible are now routinely recorded.

At the same time, plasma physics, atmospheric electricity, and complex systems research have matured to the point where speculative models can be meaningfully constrained by physical principles rather than conjecture alone.

Equally important is the cultural moment. The narrative of extraterrestrial craft has dominated discourse for decades, yet it has produced little explanatory progress. A new framework does not invalidate that narrative, but it does loosen its monopoly.

By considering the possibility of an indigenous, electromagnetic biosphere, we open conceptual space for explanations that are both less anthropocentric and more consistent with the observed data.

An Invitation Rather Than a Verdict

This book does not conclude with certainty, but with an invitation. The plasma-entity hypothesis presented here is not a final answer, but a working model. A model that can be tested, challenged, refined, or rejected as evidence accumulates.

It asks researchers, observers, and institutions to look again at familiar data through a different lens, and to consider whether the insistence on craft and visitors has obscured a subtler, and perhaps more profound reality.

If Earth does harbour a hidden biosphere composed of fields and charge, instead of cells and chemistry, then humanity is not alone in the way it has imagined, but neither is it necessarily visited in the way it has feared. Instead, it is embedded within a layered ecology that has always been present, operating at the margins of perception.

To recognise such a possibility would not diminish humanity's place in the universe, but would expand it. It would remind us that life, intelligence, and organisation are not constrained to the narrow forms we know best, and that the universe may be far more creative than our limited categories allow.

Whether this hypothesis ultimately proves correct or not, the act of asking the question remains valuable. Progress begins with the willingness to reconsider what we believe to be settled knowledge.

APPENDIX A

Testability, Constraints, and Empirical Predictions of the Plasma Entity Hypothesis

A.1

Introduction: The Role of Testability

Any hypothesis proposing novel forms of organised complexity must ultimately be assessed not by narrative coherence or explanatory breadth alone, but by its capacity to generate testable predictions and to be constrained or falsified by observation.

The plasma entity hypothesis (PEH) does not claim empirical confirmation; rather, it proposes that a subset of anomalous atmospheric and electromagnetic phenomena may be more coherently explained when framed as organised plasma-based systems rather than as discrete mechanical objects or misidentified conventional effects.

The purpose of this appendix is to delineate the empirical boundaries of the hypothesis, clarify where it is vulnerable to disconfirmation, and identify observations that would either strengthen or weaken its plausibility. In doing so, the PEH is presented not as a closed explanatory system, but as a provisional research framework.

A.2

Ontological Constraints

The plasma entity hypothesis is constrained by known physical laws. It does not invoke violations of conservation principles, exotic matter, or undiscovered forces. Instead, it operates entirely within established domains of plasma physics, electromagnetism, atmospheric electricity, and complex systems theory, albeit at scales and configurations that remain underexplored.

Accordingly, the hypothesis is bounded by the following constraints:

1. Energy availability: Any persistent plasma-based structure must be energetically sustained by identifiable environmental sources such as atmospheric electric fields, geomagnetic coupling, solar input, or ground–atmosphere charge gradients.

2. Environmental dependence: Plasma entities, if they exist, must preferentially manifest in regions conducive to plasma stability, including but not limited to storm systems, auroral zones, mineral-rich terrains, seismic fault regions, and areas of strong electromagnetic activity.

3. Finite persistence: No claim is made for indefinite stability. Plasma-based systems are expected to exhibit lifetimes constrained by energy input, environmental conditions, and field coherence.

4. Non-universality: The hypothesis does not propose that all unidentified anomalous phenomena are plasma-based. It applies only to a subset exhibiting specific behavioural and physical signatures.

A.3

Falsifiability Criteria

The PEH is falsifiable in several concrete ways. It would be significantly weakened, or rendered untenable, if any of the following conditions were consistently demonstrated:

1. Absence of electromagnetic signatures: If phenomena attributed to plasma entities repeatedly exhibit no measurable electromagnetic emissions, ionisation effects, or field interactions across multiple sensing modalities, the plasma interpretation would lack physical grounding.

2. Rigid object behaviour: Persistent demonstration that anomalous phenomena behave as rigid, inertia-bound solids under all observed conditions, including fragmentation producing conventional debris, would favour engineered-object explanations over plasma-based organisation.

3. Independence from environmental conditions: If such phenomena occur with equal frequency and stability in environments demonstrably hostile to plasma

formation (e.g. low-field, low-ionisation conditions), the hypothesis would be undermined.

4. Failure under controlled perturbation: If directed electromagnetic probing consistently destroys or irreversibly destabilises these phenomena in a manner indistinguishable from conventional plasma decay, claims of organised coherence would require re-evaluation.

A.4

Empirical Predictions

The plasma entity hypothesis makes a number of specific, testable predictions that distinguish it from craft-based or purely psychological models.

Plasma-based systems are expected to exhibit:

- Broadband electromagnetic emissions, often non-thermal in character.
- Variable optical spectra influenced by ion species, charge density, and environmental coupling.
- Frequency-dependent visibility, with some structures observable only in non-visible bands.

Prediction: multi-spectral imaging should reveal structures or behaviours invisible to the naked eye but correlated across electromagnetic bands.

A. 4.1

Behavioural Response to Electromagnetic Interaction

If organised plasma systems exist, they should respond dynamically to external electromagnetic stimuli.

Prediction:

- Radar illumination, laser exposure, or strong RF fields should induce changes in trajectory, luminosity, or configuration without necessarily causing destruction.
- Responses should vary with frequency, power, and modulation pattern.

Such behaviour would be difficult to reconcile with passive atmospheric effects.

A. 4.2

Fragmentation Without Mechanical Failure

Distributed plasma coherence implies that disruption may produce division rather than catastrophic failure.

Prediction:

- Partial fragmentation may result in multiple coherent substructures rather than debris.
- Post-fragmentation motion should remain non-ballistic, and field coupled.

This prediction differentiates plasma systems from mechanical craft.

A. 4.3

Environmental Correlations

Plasma entities should exhibit statistically significant correlations with:

- Thunderstorm activity
- Geomagnetic disturbance
- Mineralised or conductive ground substrates
- Atmospheric inversion layers

Prediction: long-term data aggregation should reveal non-random clustering tied to electromagnetic ecology rather than population density or air traffic.

A.5

Experimental and Observational Approaches

The hypothesis suggests several avenues for empirical investigation without requiring exotic instrumentation:

1. Multi-band sensor arrays combining optical, infrared, ultraviolet, RF, and magnetometric data.
2. Controlled electromagnetic probing using low-power, frequency-swept emissions to assess responsiveness.
3. Long-duration environmental monitoring in known hotspot regions.

4. Time-resolved spectroscopy during transient luminous events.

Crucially, these approaches allow discrimination between plasma-based, mechanical, and psychological explanations.

A.6

Ethical and Methodological Considerations

If anomalous plasma systems exhibit organised, adaptive behaviour, then aggressive interference may bias results or induce artefactual responses. From a methodological standpoint, observation should precede perturbation, and perturbation should be incremental and reversible where possible.

This consideration is not an ethical assertion but a practical one: complex systems are best understood under minimal disturbance.

A. 7

Status of the Hypothesis

The plasma entity hypothesis remains speculative. It does not claim verification, nor does it seek exemption from standard scientific scrutiny. Its value lies in its capacity to unify disparate observations under a physically admissible framework and to generate predictions that are distinguishable from competing models.

Whether the hypothesis ultimately proves correct, partially correct, or incorrect, it serves a constructive function by expanding the conceptual space within which anomalous phenomena may be examined.

A.8

Concluding Remarks

Scientific progress rarely arises from certainty. It emerges from the careful exploration of anomalies, the disciplined testing of unconventional ideas, and the willingness to revise or abandon hypotheses in light of evidence.

The plasma entity hypothesis invites such scrutiny. It neither demands acceptance nor resists falsification. Progress begins with the willingness to reconsider what we believe to be settled knowledge.

APPENDIX B

SUMMARY OF EMPIRICAL PREDICTIONS OF THE PLASMA ENTITY HYPOTHESIS (PEH)

Table B.1

Testable predictions derived from the Plasma Entity Hypothesis (PEH)

Domain	Prediction (Plasma Entity Hypothesis)	Conventional Expectation	Test or Observation	Key Discriminating Indicator	Relevant Chapters
Optical signature	Luminous non-solid forms with soft boundaries and internal structure	Solid craft or point light sources	Multi-spectral imaging (visible, IR, UV, NIR)	Wavelength-dependent morphology and partial transparency	1; 6; 7; 14
Motion and kinematics	Abrupt acceleration without inertial effects	Continuous acceleration limited by inertia	High-frame-rate video analysis	Instantaneous vector changes without sonic boom	6; 7; 13; 14
Radar interaction	Intermittent or fluctuating radar returns	Stable radar cross-section	Multi-band radar	Radar return varies with EM conditions	12; 21
Response to EM probing	Behavioural change under radar or RF illumination	No response beyond sensor effects	Active radar or RF exposure	Trajectory or luminosity changes without heating	12; 21
Fragmentation behaviour	Partial fragmentation without loss of coherence	Catastrophic failure on impact	Kinetic impact or near events	Persistence or division rather than destruction	14; 21
Environmental correlation	Clustering near storms, faults	No systematic environmental correlation or mineral-rich regions	Geological and meteorological mapping	Repeatability tied to EM or geological conditions	3; 5; 6; 23
Acoustic phenomena	Low-frequency or infrasonic emissions linked to geometry	Mechanical or atmospheric noise	Audio spectral analysis and resonance modelling	Frequency–size scaling consistent with cavity resonance	18
Physical traces	Non-mechanical plant deformation and node expansion	Crushing or breakage from contact	Crop and vegetation analysis	Microwave-like effects without mechanical damage	19
Human perceptual effects	Cognitive or emotional entrainment via EM coupling	Stress or misperception only	Physiological or EEG monitoring	Correlated neural effects without trauma	17; 22
Persistence over time	Long-term recurrence at fixed locations	Random or one-off events	Longitudinal site monitoring	Decades-long stability of phenomena	6; 12; 21; 23

Glossary

The following glossary is provided to clarify key terms and concepts used throughout this book. Many of these terms draw upon established principles from plasma physics, electromagnetism, neuroscience, and atmospheric science, while others are introduced here as part of the plasma entity hypothesis (PEH) and the broader theoretical framework developed in these chapters.

Where possible, definitions are expressed in neutral, descriptive language, clearly distinguishing empirically established phenomena from speculative interpretations. The glossary is intended not to persuade, but to assist the reader in navigating an interdisciplinary vocabulary that spans physics, biology, cognition, and cultural interpretation.

Terms are defined as they are used within the context of this work and may therefore differ in emphasis from their usage in other disciplines.

Behavioural residue

A persistent physical, energetic, or structural trace produced as a by-product of an entity's movement, interaction, or internal dynamics, rather than as an intentional signal or communication.

In this work, the term is used to describe environmental imprints (such as crop formations, acoustic signatures, or field disturbances) resulting from the behaviour of plasma-based or non-biological systems.

Boundary-Crossing

The process by which plasma entities interact across perceptual, physical, cognitive, or emotional thresholds separating human experience from non-biological electromagnetic intelligences. Boundary-crossing may occur through resonance, proximity, or transient environmental conditions rather than physical traversal in the conventional sense.

Cognitive Entrainment

A process by which neural oscillations in the human brain become synchronised with external electromagnetic fields. Within the Plasma Entity Hypothesis, cognitive entrainment provides a potential mechanism for subtle influence, intuition, emotional shifts, or perceived communication without direct physical contact.

Electromagnetic Biosphere

A proposed parallel ecological domain composed of plasma-based or field-structured lifeforms inhabiting atmospheric, ionospheric, and near-Earth electromagnetic environments.

Unlike the biological biosphere, this domain operates primarily through charge, resonance, and field coherence rather than cellular metabolism.

Emergent systems

Systems in which coherent structure, organisation, or behaviour arises from local interactions among components, rather than from centralised control or predefined design.

Emergent systems may exhibit stability, adaptability, and complex behaviour disproportionate to their constituent elements. In this work, the term is applied to plasma-based and electromagnetic phenomena capable of self-organisation under suitable environmental conditions.

Ethological

Referring to the study or interpretation of observable behaviour as an expression of organisation, responsiveness, or agency, without implying biological composition, cellular structure, or genetic processes.

Used here to describe patterned actions or behavioural residues of plasma-based or non-biological systems, analogous to ethology in biological organisms but not restricted to animal life or biological systems.

See also: Behavioural residue; Emergent systems; Plasma Entity Hypothesis (PEH)

Field Entanglement

As used in this work, field entanglement refers to sustained electromagnetic coupling, phase correlation, or resonance between structured fields across nested or interacting domains.

This usage does not imply nonlocal quantum entanglement, but rather coherent classical and mesoscopic field interactions capable of correlated behaviour.

Plasma Condensation (Biological Analogy)

A transient state in which a plasma entity increases local charge density, coherence, or luminosity, appearing more structured or locally physically influential.

This is not condensation in the thermodynamic sense, but an analogy describing a shift toward greater field organisation or interaction with matter.

Plasma Entity

A hypothetical non-biological lifeform composed primarily of coherent plasma and electromagnetic field structures. Such entities may exhibit organisation, responsiveness, memory, and adaptive behaviour without reliance on carbon-based biology or solid physical form.

Plasma Entity Hypothesis (PEH)

The central theoretical framework of this book, proposing that many anomalous aerial, atmospheric, and luminous phenomena may be best understood as manifestations of plasma-based lifeforms rather than technological craft, hallucination, or purely meteorological effects.

Resonant Cavity - Plasma Context

A geometric or field-defined region in which electromagnetic oscillations are reinforced through constructive interference. In plasma systems, resonant cavities may stabilise charge distributions, store energy, or support coherent field behaviour analogous to information processing structures.

Resonance

The amplification or stabilisation of oscillatory behaviour when frequencies align. Resonance is a foundational concept in physics and is used here to describe how plasma entities may organise internally, interact with their environment, or couple with biological systems such as the human brain.

Self-organisation

The spontaneous emergence of ordered structure or behaviour from local interactions within a system, without external control or central coordination. In plasma and electromagnetic systems, self-organisation may arise through feedback, resonance, and field coupling.

Storm Body

A term used to describe a large-scale storm system exhibiting unusual coherence, persistence, or internal structuring suggestive of organised plasma behaviour. Within the PEH, some storms may function as habitats for plasma entities or represent plasma entities themselves operating at meteorological scale.

Thought Influence

A subtle form of interaction in which an external electromagnetic structure may bias cognition, emotion, or decision-making without overt sensory input. Thought influence does not imply control, but modulation within the bounds of human psychological and neurological variability.

Threshold States

Conditions occurring at the boundary between waking and sleep, visibility and invisibility, coherence and noise, or stability and transition. Plasma entities are hypothesised to interact most readily during such states, where perception and electromagnetic sensitivity may be heightened.

Transient Visibility

The brief appearance of a plasma entity within the visible spectrum due to changes in charge density, ionisation, or environmental coupling. Transient visibility accounts for the fleeting, polymorphic, and often ambiguous nature of many reported sightings.

References

PREFACE

Chen, F. F. (2016). Introduction to Plasma Physics and Controlled Fusion (3rd ed.). Springer.

Hecht, E. (2017). Optics (5th ed.). Pearson.

SIONYX, LLC. (2020). SiOnyx Aurora Pro – User Manual & Specifications (Full-spectrum 400–1100 nm).

Canon Inc. (2006). EOS 400D (Digital Rebel XTi) – Instruction Manual & Specifications.

Smith, R. L., & Wagner, N. (2018). Near-Infrared (NIR) Imaging Fundamentals and Applications. Elsevier.

INTRODUCTION

Alfvén, H. (1981). Cosmic Plasma. D. Reidel.

Bellan, P. M. (2006). Fundamentals of Plasma Physics. Cambridge University Press.

Chen, F. F. (2016). Introduction to Plasma Physics and Controlled Fusion (3rd ed.). Springer.

Conselice, C. J., Wilkinson, A., Duncan, K., & Mortlock, A. (2016). The evolution of galaxy number density at $z < 8$ and its implications. The Astrophysical Journal, 830(2), 83.

NASA. (n.d.). How many stars are in the Milky Way? NASA/Caltech outreach pages (est. 1–4×10^{11}).

Wilson, T. D. (2002). Strangers to Ourselves: Discovering the Adaptive Unconscious. Harvard University Press.

Zimmermann, H. J. (1989). Neurological Limitations of Human Information Processing (classic estimates of perceptual bandwidth). Springer.

PART ONE

1 UFOs, UAPs and Other Lifeforms

Scientific and Academic Sources

AARO (All-domain Anomaly Resolution Office). (2023–2025). UAP reports, definitions, and public FAQs. U.S. Department of Defence.

Office of the Director of National Intelligence (ODNI). (2021, June 25). Preliminary Assessment: Unidentified anomalous phenomena.

U.S. Department of Defence. (2020, April 27). Statement on the release of three Navy videos (FLIR1/"Tic Tac" [2004], GIMBAL [2015], GOFAST [2015]).

Knuth, K. H., Powell, M. J., & Reali, G. (2019). Estimating Flight Characteristics of Anomalous Unidentified Aerial Vehicles. Entropy, 21(10), 939.

Ruppelt, E. J. (1956). The Report on Unidentified Flying Objects. Doubleday.

Cultural and Historical Sources

Vallée, J., & Vallée, P. (2017). Wonders in the Sky. Anomalist Books.

Graves, R., Fravor, D., & Dietrich, A. (2023). Congressional testimony on UAP observations (U.S. House Oversight Committee).

Elizondo, L. (2018–2020). Public statements outlining the "Five Observables". To The Stars Academy / media transcripts.

Cooper, H., Blumenthal, R., & Kean, L. (2017–2019). New York Times reports on UAP sightings.

2 Movement of a Species

Scientific and Academic Sources

Buehler, R., 2004. *Ionic Wind Propulsion: Electrostatic Propulsion Without Moving Parts.* NASA Technical Memorandum 213554.

Christensen, M. and Ko, T., 2018. Electroaerodynamic propulsion for sustained flight. *Nature,* 563(7732), pp.532–535.

Constable, T.J., 1976. *The Cosmic Pulse of Life.* Borderland Sciences Research Foundation.

Cook, N., 2002. *The Hunt for Zero Point: Inside the Classified World of Antigravity Technology.* Broadway Books.

Faraday, M., 1846. Experimental researchs in electricity. *Philosophical Transactions of the Royal Society of London,* 136, pp.1–20.

Frolov, A., 2004. Electrogravitics and electrokinetic propulsion. *Journal of New Energy,* 7(3), pp.68–74.

Lieberman, M.A. and Lichtenberg, A.J., 2005. *Principles of Plasma Discharges and Materials Processing.* 2nd ed. Wiley-Interscience.

Popovic, Z., 2006. Electromagnetic propulsion and applications in plasma physics. *Annual Review of Fluid Mechanics,* 38, pp.27–49.

Smirnov, B.M., 2001. *Physics of Ionised Gases.* Wiley-VCH.

Smolin, L., 2019. *Einstein's Unfinished Revolution: The Search for What Lies Beyond the Quantum.* Penguin.

Stenzel, R.L., 1999. Magnetic structures in collisionless plasmas. *Physics of Plasmas,* 6(11), pp.4450–4456.

Tajmar, M., 2014. *Advanced Space Propulsion Systems.* Springer.

Temple, R., 2022. *A New Science of Heaven.* Coronet.

Thomas, C.A., et al., 2022. Electrohydrodynamic thrusters for low-speed flight: A review. *IEEE Transactions on Plasma Science,* 50(4), pp.782–793.

Thomas, S.R., 2003. The ion-propelled aircraft: A new concept. *Journal of Electrostatics,* 60(1), pp.57–67.

Wesson, J., 2011. *Tokamaks.* 4th ed. Oxford University Press.

Wolfram, S., 2021. *A New Kind of Science.* 10th Anniversary Edition. Wolfram Media.

3 Intelligence Beyond the Veil of Perception

Alfvén, H. (1981). Cosmic Plasma. D. Reidel.

Bellan, P. M. (2006). Fundamentals of Plasma Physics. Cambridge University Press.

Bonitz, M., Henning, C., & Block, D. (2010). Complex plasmas: A laboratory for strong correlations. Reports on Progress in Physics, 73(6), 066501.

Bousso, R. (2002). The holographic principle. Reviews of Modern Physics, 74(3), 825–874.

Chen, F. F. (2016). Introduction to Plasma Physics and Controlled Fusion (3rd ed.). Springer.

Eberhard, W. (1986). A Dictionary of Chinese Symbols: Hidden Symbols in Chinese Life and Thought. Routledge. (Dragons in Chinese cosmology.)

Fortov, V. E., Khrapak, S. A., Khrapak, A. G., Molotkov, V. I., & Petrov, O. F. (2004). Dusty plasmas. Physics–Uspekhi, 47(5), 447–492.

Green, M. B., Schwarz, J. H., & Witten, E. (1987). Superstring Theory (Vols. 1–2). Cambridge University Press.

Haken, H. (1983). Synergetics: An Introduction (3rd ed.). Springer.

Laughlin, R. B. (2005). A Different Universe: Reinventing Physics from the Bottom Down. Basic Books.

Maldacena, J. (1998). The large-N limit of superconformal field theories and supergravity. Advances in Theoretical and Mathematical Physics, 2, 231–252.

Morfill, G. E., & Ivlev, A. V. (2009). Complex plasmas: An interdisciplinary research field. Reviews of Modern Physics, 81(4), 1353–1404.

Morowitz, H. J., & Sagan, C. (1967). Life in the clouds of Venus? Nature, 215, 1259–1260.

Polchinski, J. (1998). String Theory (Vols. 1–2). Cambridge University Press.

Prigogine, I., & Stengers, I. (1984). Order out of Chaos. Heinemann.

Randall, L., & Sundrum, R. (1999). An alternative to compactification. Physical Review Letters, 83(23), 4690–4693.

Shukla, P. K., & Mamun, A. A. (2002). Introduction to Dusty Plasma Physics. Institute of Physics Publishing.

Shukla, P. K., & Eliasson, B. (2009). Formation and dynamics of dust acoustic solitons in plasmas. Physics Letters A, 373(35), 3165–3168.

Susskind, L. (1995). The world as a hologram. Journal of Mathematical Physics, 36(11), 6377–6396.

Thomas, H. M., Morfill, G. E., Demmel, V., Goree, J.,

Feuerbacher, B., & Möhlmann, D. (1994). Plasma crystal: Coulomb crystallization in a dusty plasma. Physical Review Letters, 73(5), 652–655.

’t Hooft, G. (1993). Dimensional reduction in quantum gravity. Salamfestschrift, World Scientific, 284–296.

Vallée, J., & Vallée, P. (2017). Wonders in the Sky (updated ed.). Anomalist Books. (Historical luminous phenomena.)

Zwiebach, B. (2009). A First Course in String Theory (2nd ed.). Cambridge University Press.

Chester, K. (2007). Strange Company: Military Encounters with UFOs in World War II. Anomalist Books. (Foo fighters.)

Dahl, R. (1943). The Gremlins. Random House. (Cultural origin of “gremlins”).

4 The Plasma Interface of Perception

Scientific and Academic Sources

Alfvén, H. (1981). Cosmic Plasma. D. Reidel.

Bellan, P. M. (2006). Fundamentals of Plasma Physics. Cambridge University Press.

Bonitz, M., et al. (2010). Complex plasmas. Reports on Progress in Physics, 73(6), 066501.

Fortov, V. E., et al. (2004). Dusty plasmas. Physics–Uspekhi, 47(5), 447–492.

McDonnell, M. D., et al. (2008). Stochastic Resonance. Cambridge University Press.

Morfill, G. E., & Ivlev, A. V. (2009). Complex plasmas. Reviews of Modern Physics, 81(4), 1353–1404.

Prigogine, I., & Stengers, I. (1984). Order out of Chaos. Heinemann.

Shukla, P. K., & Mamun, A. A. (2002). Introduction to Dusty

Plasma Physics. Institute of Physics Publishing.

Susskind, L. (1995). The world as a hologram. Journal of Mathematical Physics, 36(11), 6377–6396.

Wolfram, S. (2021). A New Kind of Science (2nd ed.). Wolfram Media.

Wolfram, S. (2023). The Physicalisation of Meta-Mathematics. Wolfram Research blog.

Cultural and Historical Sources

Yaden, D. B., & Newberg, A. B. (2022). The Varieties of Spiritual Experience. Oxford University Press. (Altered states and luminous entities.)

Atmanspacher, H., & Fach, W. (2019). Exceptional experiences of stable selfhood. Frontiers in Psychology, 10, 1630.

5 Cosmology at the Event Horizon

Scientific and Academic Sources

Arkani-Hamed, N., Dimopoulos, S., & Dvali, G. (1998). The hierarchy problem and new dimensions. Physics Letters B, 429(3–4), 263–272.

Bousso, R. (2002). The holographic principle. Reviews of Modern Physics, 74(3), 825–874.

Greene, B. (2000). The Elegant Universe. W. W. Norton.

Maldacena, J. (1998). The large-N limit of superconformal field theories. Advances in Theoretical and Mathematical Physics, 2, 231–252.

NASA/ESA/CSA (2023–2024). James Webb Space Telescope early release results. STScI.

Natarajan, P. (2024). Mapping the Heavens. Yale University Press.

Popławski, N. J. (2010). Cosmology with torsion. Physics Letters B, 694(3), 181–185.

Popławski, N. J. (2012). Universe in a Black Hole. Scientific American, July 2012.

Randall, L., & Sundrum, R. (1999). An alternative to compactification. Physical Review Letters, 83(23), 4690–4693.

Smolin, L. (1997). The Life of the Cosmos. Oxford University Press.

Susskind, L. (2005). The Cosmic Landscape. Little, Brown.

't Hooft, G. (1993). Dimensional reduction in quantum gravity. Salamfestschrift.

Witten, E. (1996). Bound states of strings and p-branes. Nuclear Physics B, 460(2–3), 335–350.

Zwicky, F. (1937). On the masses of nebulae. The Astrophysical Journal, 86, 217.

Cultural and Historical Sources

Carroll, S. (2010). From Eternity to Here. Dutton. (Bridges physics with philosophical implications of time.)

Susskind, L. (2005). The Cosmic Landscape. Little, Brown. (Borders science and broader cosmological speculation.)

6 The Liminal Veil

Scientific and Academic Sources

Alfvén, H. (1981). Cosmic Plasma. D. Reidel.

Baumjohann, W., & Treumann, R. A. (1997). Basic Space Plasma Physics. Imperial College Press.

Blinov, D., et al. (2018). Polarimetric detection of Kordylewski dust clouds. Monthly Notices of the Royal Astronomical Society, 479(3), 3537–3544.

Fortov, V. E., Ivlev, A. V., Khrapak, S. A., Khrapak, A. G., &

Morfill, G. E. (2005). Complex (dusty) plasmas: Current status, open issues, perspectives. Physics Reports, 421(1–2), 1–103.

Horányi, M., Sternovsky, Z., Szalay, J., & Kempf, S. (2015). A permanent, asymmetric dust cloud around the Moon. Nature, 522, 324–326.

Josephson, B. D. (1962). Possible new effects in superconductive tunnelling. Physics Letters, 1(7), 251–253.

Kordylewski, K. (1961). Photographic evidence for dust accumulation at the lunar triangular points of equilibrium. Acta Astronomica, 11, 165–170.

Morfill, G. E., & Ivlev, A. V. (2009). Complex plasmas: An interdisciplinary research field. Reviews of Modern Physics, 81(4), 1353–1404.

NASA (2024, May 10–11). Reports on the May 2024 G5 geomagnetic storm. Space Weather Prediction Centre (NOAA) & European Space Agency bulletins.

Popławski, N. J. (2010). Cosmology with torsion: An alternative to cosmic inflation. Physics Letters B, 694(3), 181–185.

Prigogine, I., & Stengers, I. (1984). Order out of Chaos. Heinemann.

Randall, L., & Sundrum, R. (1999). An alternative to compactification. Physical Review Letters, 83(23), 4690–4693.

Schumann, W. O. (1952). On the free oscillations of a conducting sphere which is surrounded by an air layer and an ionosphere shell. Zeitschrift für Naturforschung A, 7(2), 149–154.

Smolin, L. (1997). The Life of the Cosmos. Oxford University Press.

Susskind, L. (2005). The Cosmic Landscape: String Theory and the Illusion of Intelligent Design. Little, Brown.

Temple, R. (2022). A New Science of Heaven. Coronet.

Wdowczyk, J., & Wolfendale, A. W. (1977). Cosmic rays and the South Atlantic Anomaly. Nature, 268, 510–512.

Zhou, Y.- L., et al. (2024). Ionospheric and thermospheric responses to the May 2024 geomagnetic superstorm observed by Swarm and MSS-1 satellites. Journal of Geophysical Research: Space Physics, 129(8), e2024JA033212.

Biblical References

Exodus 3:2–4 (The burning bush).

Exodus 13:21 (The pillar of fire).

Ezekiel 1:15–21 (The wheels within wheels vision).

Matthew 2:1–12 (The Star of Bethlehem).

2 Kings 2:11 (Elijah taken up by fire from heaven).

Mesoamerican

Taube, K. A. (1995). The Feathered Serpent and Flowering Trees: The Iconography of Life, Death, and the Underworld in Ancient Mesoamerica. RES: Anthropology and Aesthetics, 29/30, 53–83. (On Quetzalcoatl/Itzamna.)

Chinese

Eberhard, W. (1986). A Dictionary of Chinese Symbols: Hidden Symbols in Chinese Life and Thought. Routledge. (On lóng dragons and pearl imagery.)

Norse

Simek, R. (1993). Dictionary of Northern Mythology. Boydell & Brewer. (On Bifröst and Norse cosmology.)

Vedic / Hindu

Feuerstein, G. (1996). The Shambhala Encyclopaedia of Yoga. Shambhala. (On Vimanas, Devas, and Akashic concepts.)

Aboriginal Australian

Mowaljarlai, D., & Malnic, J. (1993). Yorro Yorro: Aboriginal Creation and the Renewal of Nature. Magabala Books. (On Wandjina spirits.)

7 Spirals of Dust and Light

Scientific and Academic Sources

Balme, M. R., & Greeley, R. (2006). Dust devils on Earth and Mars. Reviews of Geophysics, 44(3), RG3003.

Carroll, J. J., & Ryan, J. A. (1970). Atmospheric vortices and dust devil studies. Journal of the Atmospheric Sciences, 27(3), 464–471.

Farrell, W. M., et al. (2004). Electric and magnetic signatures of dust devils from the 2000–2001 MATADOR desert tests. Journal of Geophysical Research: Planets, 109(E3).

Freier, G. D. (1960). The electric field of a dust devil. Journal of Geophysical Research, 65(10), 3504–3504.

Kok, J. F., et al. (2012). The physics of wind-blown sand and dust. Reports on Progress in Physics, 75(10), 106901.

Metzger, S. M., et al. (1999). Dust devils as aeolian transport on Mars. Geophysical Research Letters, 26(18), 2781–2784.

Renno, N. O., et al. (1998). Electrical fields and dust devils on Mars. Journal of Geophysical Research: Planets, 103(E1), 313–325.

Shukla, P. K., & Mamun, A. A. (2002). Introduction to Dusty Plasma Physics. Institute of Physics Publishing.

Yair, Y. (2008). Charge generation and separation processes. Space Science Reviews, 137(1–4), 119–131.

Aboriginal Dreamtime narratives (Pilbara region)

Persinger, M. A. (1987). Neuropsychological Bases of God Beliefs. Praeger. (Environmental EM fields and perception of entities.)

Temple, R. (2022). A New Science of Heaven. Watkins.

Vallée, J., & Vallée, P. (2017). Wonders in the Sky. Anomalist Books.

8 Earth's Hidden Order

Classical and Historical Sources

Euclid. (1956). The Thirteen Books of Euclid's Elements (T. L. Heath, Trans.). Dover. (Original work c. 300 BCE).

Plato. (1961). Timaeus (D. Lee, Trans.). Penguin Classics.

Plato. (1997). Republic (Book VII: The Allegory of the Cave). Hackett.

Jung, C. G. (1968). Archetypes and the Collective Unconscious. Princeton University Press.

Lawlor, R. (1982). Sacred Geometry: Philosophy and Practice. Thames & Hudson.

Geometry, Mathematics, and Physics

Coxeter, H. S. M. (1973). Regular Polytopes. Dover.

Klee, V., & Wagon, S. (1991). Old and New Unsolved Problems in Plane Geometry and Number Theory. MAA.

Arkani-Hamed, N., Bourjaily, J., Cachazo, F., Goncharov, A. B., Postnikov, A., & Trnka, J. (2016). Grassmannian Geometry of Scattering Amplitudes. Cambridge University Press.

Lisi, A. G. (2007). An exceptionally simple theory of everything. arXiv preprint arXiv:0711.0770. (E8 lattice model).

Aurich, R., Lustig, S., & Steiner, F. (2004). CMB anisotropy of the Poincaré dodecahedron. Classical and Quantum Gravity, 21(20), 4901–4925.

Natural Patterns and Biological Systems

Ball, P. (2012). Shapes: Nature's Patterns – A Tapestry in Three Parts. Oxford University Press.

Livio, M. (2002). The Golden Ratio: The Story of Phi, the World's Most Astonishing Number. Broadway Books.

Stewart, I. (2017). The Beauty of Numbers in Nature: Mathematical Patterns and Principles from the Natural World. MIT Press.

Watts, A., et al. (2010). Icosahedral virus structure and assembly. Annual Review of Virology, 1, 33–46.

Kustov, V., et al. (2008). DNA topology and dodecahedral symmetry. Biopolymers and Cell, 24(2), 119–125.

Esoteric and Consciousness Sources

Heindel, M. (1909). The Rosicrucian Cosmo-Conception. Rosicrucian Fellowship.

Blavatsky, H. P. (1888). The Secret Doctrine. Theosophical Publishing Company.

McKenna, T. (1992). Food of the Gods. Bantam. (Reports of geometric visionary states).

Strassman, R. (2001). DMT: The Spirit Molecule. Park Street Press.

9 Crystals and Planetary Lifeforms

Scientific and Academic Sources

Barlow, W. (1897). Crystallographic axes and symmetry. Journal of the Chemical Society, Transactions, 71, 254–263.

Buerger, M. J. (1956). Elementary Crystallography. Wiley.

Dana, J. D. (1997). Dana's New Mineralogy (8th ed.). Wiley.

Gaines, R. V., Skinner, H. C. W., Foord, E. E., Mason, B., & Rosenzweig, A. (1997). Dana's New Mineralogy: The System of Mineralogy of James Dwight Dana and Edward Salisbury Dana. Wiley.

Holden, E.-N., et al. (2018). Metamictisation of zircon and

implications for geochronology. Earth and Planetary Science Letters, 498, 101–113.

International Union of Crystallography (IUCr). (2022). Crystallographic Information Framework. Retrieved from www.iucr.org.

Klein, C., & Dutrow, B. (2011). Manual of Mineral Science (23rd ed.). Wiley.

Makovicky, E. (2013). Crystallography of complex sulphides. Reviews in Mineralogy and Geochemistry, 61(1), 7–46.

Pabst, A. (1981). Metamict minerals. American Mineralogist, 66(9–10), 924–945.

Plimer, I. R. (1998). A Short History of Geology. Allen & Unwin.

Rayner, J. (2015). Geology and mineral deposits of Western Australia. Geological Survey of Western Australia Bulletin, 145, 1–262.

Wilson, A. J. C. (1992). International Tables for Crystallography (Vol. A). Kluwer Academic.

Australian Geology & Regional Mineralogy

Bagas, L., Bierlein, F. P., Nelson, D. R., & McNaughton, N. J. (2004). Geology and mineralisation of the Yilgarn Craton, Western Australia. Ore Geology Reviews, 25(3–4), 247–270.

Trendall, A. F. (2002). The Hamersley Province: Genesis of banded iron formations. Geological Survey of Western Australia.

Tyler, I. M., & Hocking, R. M. (2011). The geology of Western Australia's Kimberley region. Geological Survey of Western Australia.

Speculative / Integrative Sources

Temple, R. (2022). A New Science of Heaven. Watkins. (Links

between crystalline lattices and plasma life.)

Vallee, J. (2010). Dimensions: A Casebook of Alien Contact. Anomalist Books. (Historical references to lattices and resonant structures.)

10 Geology and Plasma Lifeforms

Scientific and Academic Sources

Alfven, H. (1981) *Cosmic Plasma.* Dordrecht: Reidel.

Anderson, R. and Hamilton, R. (1965) 'Electroluminescence in Stressed Quartz', *Nature*, 205(4970), pp. 59–60.

Arora, B.R., Baldwin, R.T. and Lilley, F.E.M. (1995) 'The Geoelectric Field Associated with Earthquake Processes', *Surveys in Geophysics*, 16, pp. 159–192.

Birkeland, K. (1908) *The Norwegian Aurora Polaris Expedition 1902–1903.* Christiania: H. Aschehoug.

Blomqvist, A. (2010) 'The Hessdalen Lights: A Scientific Mystery', *Journal of Scientific Exploration*, 24(2), pp. 201–222.

Boccippio, D. et al. (1995) 'Sprites, ELF Transients, and Positive Lightning', *Science*, 269(5227), pp. 1088–1091.

Brady, B.T. and Rowell, G.A. (1986) 'Laboratory Investigation of the Electrodynamics of Rock Fracture', *Journal of Geophysical Research*, 91(B7), pp. 7495–7504.

Devereux, P. (1982) *Earth Lights: The Geological Basis of Strange Phenomena.* London: Turnstone Press.

Devereux, P. and Krippner, S. (1989) *Earth Lights Revelation.* London: Blandford.

Key ethnographic–geophysical analysis relevant to Hessdalen and Min Min phenomena.

Freund, F.T. (2002) 'Charge Generation and Propagation in Rocks', *Journal of Geodynamics*, 33, pp. 543–570.

Freund, F.T. (2011) 'Pre-earthquake Signals: Underlying Physical Processes', *Journal of Asian Earth Sciences*, 41, pp. 383–400.

Grosz, A. and Legters, L.J. (eds.) (2011) *The Hessdalen Phenomenon: New Light on Earth Lights.* Bergen: Høgskolen i Østfold.

Henderson, M. and Adams, C. (1993) *Min Min Light: The Visitor Who Never Arrives.* Sydney: Angus & Robertson.

Johansen, H. and Strand, E.P. (1984) 'Project Hessdalen 1984: Final Technical Report'. Østfold University College.

King, B. (2020) 'Electromagnetic Signatures in Luminous Atmospheric Phenomena', *Atmospheric Research*, 235, 104748.

Leighly, J.B. (1934) 'The Formation of Electrostatic Fields by Water in Rocks', *American Journal of Science*, 28(167), pp. 409–419.

Liritzis, I. and Castro, B. (2013) 'Electric Earthquakes: Electromagnetic Phenomena in the Earth's Crust', *Geophysical Processes and Models*, 2(1), pp. 1–20.

Paiva, G.S. (2021) paper on Hessdalen lights (Springer / Meteorology and Atmospheric Physics)

Persinger, M.A. and Derr, J.S. (1985) 'Geomagnetic Flux and Luminous Phenomena Over Faults', *Perceptual and Motor Skills*, 60, pp. 59–65.

Rønne, T. (2014) 'Spectroscopic Analysis of Hessdalen Atmospheric Lights', *Astrophysics and Space Science*, 350(2), pp. 677–686.

Sasmal, S. and Ananthakrishnan, S. (1970) 'Electromagnetic Emissions from Quartz and Feldspar Under Stress', *Bulletin of the Seismological Society of America*, 60(4), pp. 1401–1410.

St-Laurent, F., Derr, J. and Freund, F. (2006) 'Earthquake Lights and the p-hole Theory', *Natural Hazards and Earth System Sciences*, 6(5), pp. 639–644.

Teodorani, M. (2004) 'A Long-Term Scientific Survey of the Hessdalen Phenomenon', *Journal of Scientific Exploration*, 18(2), pp. 217–251.

Teodorani, M. (2009) *Earth Lights and Physics*. Bologna: Noesis. Covers plasma behaviour, Hessdalen data, and EM interactions.

Zhao, Y. et al. (2013) 'Piezoelectric Luminescence from Stressed Silica', *Applied Physics Letters*, 102, 051907.

11 The Lunar Mirage

Scientific and Academic Sources

Arlin, J. D., & Stubbs, T. J. (2012). Electrostatic charging of lunar dust and implications for plasma–dust interactions. Planetary and Space Science, 74(1), 173–181.

Crotts, A. P. S. (2007). Transient Lunar Phenomena: Regularity and Reality. The Astronomical Journal, 133(2), 819–832.

Colwell, J. E., Batiste, S., Horányi, M., Robertson, S., & Sture, S. (2007). Dust transport in photoelectron layers and the formation of dust ponds on Eros. Icarus, 190(1), 127–144.

Goldsmith, D. (1972). Transient Lunar Phenomena. Scientific American, 227(1), 24–33.

Grygar, J., & Grigoryan, S. (1975). Plasma phenomena on the Moon and Transient Lunar Events. Astrophysics and Space Science, 34(1), 83–102.

Horányi, M., Szalay, J. R., Kempf, S., et al. (2015). A permanent, asymmetric dust cloud around the Moon. Nature, 522, 324–326.

Kadono, T., Arakawa, M., & Genda, H. (2002). Electrostatic

charging and dust levitation on the Moon and asteroids. Advances in Space Research, 29(8), 1263–1268.

Levasseur-Regourd, A. C., Lasue, J., Hadamcik, E., & Alcouffe, G. (2007). Physical properties of cometary and interplanetary dust. Planetary and Space Science, 55(9), 1086–1095.

O'Keefe, J. A., & Urey, H. C. (1970). Transient lunar events and lunar outgassing. Science, 167(3918), 1325–1326.

Poppe, A. R., Fatemi, S., & Halekas, J. S. (2021). Solar wind interactions with the Moon. Reviews of Geophysics, 59(1), e2020RG000706.

Samsonov, D., Morfill, G. E., & Ivlev, A. (2004). Plasma crystal lattice dynamics: Wave propagation and instabilities. IEEE Transactions on Plasma Science, 32(2), 555–559.

Severny, A. B., Terekhov, O. M., & Zvereva, A. M. (1974). Observations of lunar transient events with the Crimean Astrophysical Observatory. Icarus, 22(3), 123–136.

Shevchenko, V. V. (2007). The problem of transient lunar phenomena. Solar System Research, 41(1), 20–30.

Stubbs, T. J., Vondrak, R. R., & Farrell, W. M. (2006). A dynamic fountain model for lunar dust. Advances in Space Research, 37(1), 59–66.

Tribble, A. C. (1995). The Space Environment: Implications for Space Design. Princeton University Press.

12 Symbiosis and Humanity's Hidden Light

Biophoton and Biofield Research

Popp, F. A., Li, K. H., & Gu, Q. (1992). Biophoton Emission: Experimental Background and Theoretical Approaches. Springer.

Kobayashi, M., Kikuchi, D., & Okamura, H. (2009). Imaging of ultraweak spontaneous photon emission from human body displays diurnal rhythm. Photochemistry and Photobiology, 85(3), 650–656.

Van Wijk, R., & Van Wijk, E. P. A. (2005). Multi-site recording and spectral analysis of spontaneous photon emission from human body. Electromagnetic Biology and Medicine, 24(3), 221–239.

Rubik, B. (2002). The biofield hypothesis: Its biophysical basis and role in medicine. Journal of Alternative and Complementary Medicine, 8(6), 703–717.

Mitochondrial Origins and Symbiosis

Margulis, L. (1970). Origin of Eukaryotic Cells. Yale University Press.

Lane, N. (2005). Power, Sex, Suicide: Mitochondria and the Meaning of Life. Oxford University Press.

Roger, A. J., Muñoz-Gómez, S. A., & Kamikawa, R. (2017). The origin and diversification of mitochondria. Current Biology, 27(21), R1177–R1192.

Electromagnetic Sensitivity and Human Fields

Persinger, M. A. (2009). The possible role of global geomagnetic activity in the biology of consciousness. Neuroscience Letters, 452(3), 182–187.

Kwon, H. S., & Park, J. E. (2012). Electromagnetic field effects on the nervous system. Molecular & Cellular Toxicology, 8, 249–254.

Thar, R., & Kühl, M. (2004). Propagation of electromagnetic fields in microbial mats. Applied and Environmental Microbiology, 70(2), 1033–1040.

Kirlian Photography and Aura Studies

Krippner, S., & Rubin, D. (1974). Galaxies of Life: The Human Aura in Acupuncture and Kirlian Photography. Gordon and

Breach.

Konikiewicz, L. W., & Konikiewicz, B. W. (1979). Electrophotography and Kirlian Photography: Myth and Reality. Theosophical Publishing House.

Pehek, J. O., Kyler, H. J., & Faust, D. L. (1976). Image modulation in corona discharge photography. Science, 194(4262), 263–270.

End-of-Life and Near-Death Phenomena

Greyson, B. (2003). Incidence and correlates of near-death experiences in a cardiac care unit. General Hospital Psychiatry, 25(4), 269–276.

Fenwick, P., Lovelace, H., & Brayne, S. (2007). End-of-life experiences and their implications for palliative care. Progress in Palliative Care, 15(5), 207–213.

Borjigin, J., Lee, U., & Liu, T. (2013). Surge of neurophysiological coherence and connectivity in the dying brain. Proceedings of the National Academy of Sciences, 110(35), 14432–14437.

Cultural and Cosmological Parallels

Eliade, M. (1957). Myth and Reality. Harper & Row.

Widengren, G. (1969). The Ascension of the Apostle and the Heavenly Book. Uppsala University.

Nasr, S. H. (2007). Islamic Cosmological Doctrines. Routledge.

Klein, J. (1995). The Science of the Soul: The End of Life and the Seven Heavens. Sophia Perennis.

13 Controlling the Narrative

Scientific and Academic Sources

Battelle Memorial Institute. Project Blue Book Special Report No. 14. Wright–Patterson Air Force Base: USAF, 1955.

Central Intelligence Agency. Scientific Advisory Panel on unidentified flying objects (Robertson Panel Report). Washington DC: CIA, 1953.

Condon, E. U., & University of Colorado. Scientific Study of Unidentified flying objects. Washington DC: US Government Printing Office, 1968.

Department of Defence (US). Establishment of the All-domain Anomaly Resolution Office (AARO). 2022.

French Space Agency (CNES). GEIPAN Case Files and Research Summaries. Toulouse: CNES, 2005–present.

Ministry of Defence (UK). Project Condign: Unidentified anomalous phenomena in the UK Air Defence Region. Defence Intelligence Staff, 1997–2000 (declassified 2006).

NASA. UAP Independent Study Team: Final Report. Washington DC: NASA, 2023.

Office of the Director of National Intelligence (US). Preliminary Assessment: Unidentified anomalous phenomena. Washington DC, 2021.

Ridpath, I. (2001). "The Flying Saucer Working Party." Ministry of Defence Archives Review.

Transport Canada. Project Magnet: Geomagnetic Energy Investigations. Ottawa, 1950–1954.

~~United States~~United States Air Force. Project Grudge Report. Washington DC: USAF, 1949.

~~United States~~United States Air Force. Project Blue Book Archive. US National Archives, 1947–1969.

14 Origins of Theocracy and Coverup

Charles RH (1912) The Book of Enoch. Oxford: Clarendon Press.

Nickelsburg GWE & VanderKam JC (2004) 1 Enoch: A New

Translation. Minneapolis: Fortress Press.

Knibb MA (1978) The Ethiopic Book of Enoch. Oxford: Clarendon Press.

Schäfer P (2011) The Origins of Jewish Mysticism. Tübingen: Mohr Siebeck.

Gieschen CA (1998) Angelomorphic Christology. Leiden: Brill.

Pseudo-Dionysius (1987) The Complete Works, trans. Luibheid C & Rorem P. New York: Paulist Press.

Eliade M (1958) Patterns in Comparative Religion. London: Sheed & Ward.

Campbell J (1964) The Masks of God: Occidental Mythology. New York: Viking Press.

Temple R (2022) A New Science of Heaven. London: Century.

Foucault M (1970) The Order of Things. London: Routledge.

Berger PL & Luckmann T (1966) The Social Construction of Reality. London: Allen Lane.

Ancient Sources and Cultural Accounts

VanderKam, J. (1995) *Enoch: A Man for All Generations.* Columbia: University of South Carolina Press.

Knibb, M. A. (1978) *The Ethiopic Book of Enoch: A New Edition in the Light of the Aramaic Dead Sea Fragments.* Oxford: Clarendon Press.

Eliade, M. (1964) *Shamanism: Archaic Techniques of Ecstasy.* Princeton: Princeton University Press.

Naydler, J. (1996) *Temple of the Cosmos: The Ancient Egyptian Experience of the Sacred.* Rochester: Inner Traditions.

Briggs, K. (1976) *A Dictionary of Fairies.* London: Penguin Books.

15 ENTITIES AND AIRCRAFT

Gremlins and Aviation Folklore

HistoryNet. (2019). 'Gremlins: A pilot's worst nightmare'. *HistoryNet*, 15 October.

War History Online. (2021). 'Gremlins were a pilot's worst enemy during World War II'. *War History Online*, 1 September.

'Gremlin'. (2024). *Wikipedia, The Free Encyclopaedia.*

Foo-Fighters& WWII Aerial Anomalies

Time Magazine. (1945). 'Foo-Fighter'. *Time*, 15 January.

Smithsonian Air & Space. (2015). 'What were the mysterious "Foo fighters" sighted by WWII night flyers?'. *Air & Space Magazine.* Smithsonian Institution.

History.com. (2018). 'Mysterious UFOs seen by WWII airmen still unexplained'. *History.com*, 15 August.

Project 1947. (n.d.). 'UFO reports and official documents 1945'. *Project1947.com.*

Early Radar, "Angels & Atmospheric Echoes

'Radar angels'. (2024). *Wikipedia, The Free Encyclopaedia.*

Martin, W. J. & Shapiro, A. (2007). 'Discrimination of bird and insect radar echoes in clear air using the WSR-88D'. *Journal of Atmospheric and Oceanic Technology*, 24(7), pp. 1215–1230.

O'Haver, K. W. et al. (2018). 'Radar development for air and missile defence'. *Johns Hopkins APL Technical Digest*, 34(2), pp. 133–149.

Aircraft Electromagnetic Emissions & Environment

Civil Aviation Safety Authority. (2021). *Advisory Circular AC 21-53: Electromagnetic Compatibility.* CASA, Canberra.

Clarke, C. A. (1987). *Aircraft Electromagnetic Compatibility*. NASA Technical Memorandum 87767. National Aeronautics and Space Administration.

Singh, A. & Raj, P. (2018). 'EMI effects in flight control systems and their mitigations'. In: *Electromagnetic Compatibility of Systems*, pp. 151–180. (Chapter on EMI/EMC in flight control systems).

RTCA standards, DO-160 Environmental Conditions and Test Procedures for Airborne Equipment

Modern Radar: AESA, Passive & Photonic Radar

Leonardo UK. (2023). 'An introduction to digital AESA radars'. *Leonardo News & Stories.*

Northrop Grumman. (2024). 'Active electronically scanned array (AESA) radars'. *Northrop Grumman Mission Solutions.*

Zhang, Q. et al. (2019). 'Application overview and development trend of passive radar'. *The Journal of Engineering*, 2019(9), pp. 5829–5834.

Rai, P. K. et al. (2021). 'LTE-based passive radars and applications: A review'. *International Journal of Remote Sensing*, 42(18), pp. 6990–7016.

Sharma, A. et al. (2021). 'A cost-effective photonic radar under adverse weather conditions'. *Frontiers in Physics*, 9, 747598.

Zhang, Z. et al. (2023). 'Photonic radar for contactless vital sign detection'. *Nature Photonics*, 17, pp. 1030–1037.

(New Atlas. 2022. 'Advanced photonic radar captures images down to the centimetre scale'. *New Atlas*, 7 February.)

Modern UAP Encounters

'Pentagon UFO videos'. (2024). *Wikipedia, The Free Encyclopaedia.* Rogoway, T. (2018). 'Detailed official report on harrowing encounter between F/A-18s and UFO surfaces'. *The War Zone*, 22 May.

National Defense University Press. (2023). 'Cutting the chaff: Overlooked lessons of military UAP sightings for the joint force'. *Joint Force Quarterly*, 110, pp. 100–110.

CBS News. (2021). 'Navy pilots recall "unsettling" 2004 UAP sighting – 60 Minutes'. *CBS News*, 16 May.

Biodiversity & Unknown Species

Locey, K.J. and Lennon, J.T. (2016) 'Scaling laws predict global microbial diversity', *Proceedings of the National Academy of Sciences of the ~~United States~~United States of America*, 113(21), pp. 5970–5975.

Mora, C. et al. (2011). 'How many species are there on Earth and in the ocean?'. *PLoS Biology*, 9(8), e1001127.

Sweetlove, L. (2011). 'Number of species on Earth tagged at 8.7 million'. *Nature News*, 23 August.

Watson, T. (2011). '86 percent of Earth's species still unknown?'. *National Geographic News*, 25 August.

16 Food of the Gods

UAP and Nuclear Facility Reports

Hastings, R. (2008) *UFOs and Nukes: Extraordinary Encounters at Nuclear Weapons Sites.* Authorhouse.

Hastings, R. and Salas, R. (2010) *Faded Giant: The 1967 Missile/UFO Incidents and Subsequent Air Force Cover-Up.* CreateSpace.

Sokolov, A. (1997) 'Soviet-era observations of Unidentified anomalous phenomena', *Journal of Scientific Exploration*, 11(1), pp. 1–24.

Jacobs, D. M. (2017) *The UFO Controversy in America.* Bloomington: Indiana University Press.

Good, T. (1998) *Need to Know: UFOs, Military and Intelligence.* London: Sidgwick & Jackson.

Plasma Phenomena and Atmospheric Luminous Events

Stenhoff, M. (1999) *Ball Lightning: An Unsolved Problem in Atmospheric Physics.* New York: Kluwer Academic/Plenum.

Barry, J. D. (1980) 'The nature of ball lightning', *Nature*, 284, pp. 575–576.

Teodorani, M. (2004) 'A physical study of anomalous luminous phenomena in the Hessdalen valley', *Journal of Scientific Exploration*, 18(2), pp. 217–251.

Paiva, G. S., Oliveira, A. S. and Pasko, V. P. (2007) 'Biophoton emission from ionised air plasma', *Plasma Science and Technology*, 9(4), pp. 551–560.

Uman, M. A. (2001) *The Lightning Discharge.* New York: Dover Publications.

Electromagnetic and Environmental Interactions

Persinger, M. A. (1981) *Neuropsychological Bases of God Beliefs.* New York: Praeger.

Persinger, M. A. and Lafrenière, G. F. (1977) *Space-Time Transients and Unusual Events.* Chicago: Nelson Hall.

Devereux, P. (1989) *Earth Lights Revelation.* London: Blandford.

Vallee, J. (1990) *Confrontations: A Scientist's Search for Alien Contact.* New York: Ballantine.

Vallee, J. (1999) *Passport to Magonia: From Folklore to Flying Saucers.* London: Richard Dolan Press.

Interpretive and Comparative Works

Lewis-Williams, J. D. (2002) *The Mind in the Cave: Consciousness and the Origins of Art.* London: Thames & Hudson.

Harpur, P. (1994) *Daimonic Reality: A Field Guide to the*

Otherworld. London: Aeon Books.

Strieber, W. (1987) *Communion.* New York: Morrow.

Clark, J. B. (1998) *High Strangeness: Hyperdimensions and the Process of Contact.* Seattle: Arcturus Books.

Relevant Contextual Studies

Kallenbach, R. (2000) 'Charged particle dynamics in near-Earth space', *Reviews of Geophysics*, 38(2), pp. 295–316.

Versteegh, G. J. M. (1969) 'Plasma behaviour in the ionosphere', *Space Science Reviews*, 10(1), pp. 111–158.

Barnard, A. (2005) *The Supernatural: Anthropological Perspectives.* Cambridge: Cambridge University Press.

17 Lights over Fukushima

United States Geological Survey (2011) *M 9.0 – Near the East Coast of Honshu, Japan.* Earthquake Hazards Program. Reston, VA: U.S. Geological Survey.

Japan Meteorological Agency (2011) *The 2011 off the Pacific coast of Tōhoku Earthquake.* Tokyo: Japan Meteorological Agency.

National Police Agency of Japan (2012) *Damage Situation and Police Countermeasures Associated with the 2011 Tōhoku District – Off the Pacific Ocean Earthquake.* Tokyo: National Police Agency of Japan.

Tokyo Electric Power Company (2011–2013) *Fukushima Daiichi Nuclear Power Station Accident Overview and Analysis.* Tokyo: TEPCO.

International Atomic Energy Agency (2015) *The Fukushima Daiichi Accident.* Vienna: International Atomic Energy Agency.

World Health Organization (2013) *Health Risk Assessment from the Nuclear Accident after the 2011 Great East Japan Earthquake and Tsunami.* Geneva: World Health Organization.

United Nations Scientific Committee on the Effects of Atomic Radiation (2014) *Sources, Effects and Risks of Ionizing Radiation.* UNSCEAR 2013 Report to the General Assembly. New York: United Nations.

Knoll, G.F. (2010) *Radiation Detection and Measurement.* 4th edition. Hoboken, NJ: John Wiley & Sons.

International Commission on Radiological Protection (2007) *The 2007 Recommendations of the International Commission on Radiological Protection.* ICRP Publication 103. Oxford: Elsevier.

Mutual UFO Network (2011) *Japan Fukushima UFO Sighting Reports.* Cincinnati, OH: Mutual UFO Network.

Huffington Post (2011) *UFO Sightings Reported Near Fukushima Nuclear Plant.* New York: The Huffington Post.

The Independent (2011) *UFO Sightings Reported Over Fukushima Nuclear Facility.* London: Independent Digital News and Media.

Netflix (2023) *Encounters,* Episode 4: "Lights Over Fukushima". Directed by Yon Motskin. Los Gatos, CA: Netflix.

Fukushima Minpō (2021) *International UFO Laboratory Opens in Iinomachi.* Fukushima: Fukushima Minpō Publishing Co.

NHK World Japan (2021) *Fukushima Town Opens UFO Research Facility.* Tokyo: Japan Broadcasting Corporation.

Salas, R.L. (2010) *Faded Giant: The UFO Mystery at Malmstrom Air Force Base.* Raleigh, NC: Lulu Press.

Hastings, R. (2017) *UFOs and Nukes: Extraordinary Encounters at Nuclear Weapons Sites.* Second edition. Charleston, SC: History Press.

United States Department of Defense (2023) *Historical Record Report: Unidentified Anomalous Phenomena.* Washington, DC: U.S. Department of Defense.

The Book of Enoch (c. 3rd–1st century BCE) *1 Enoch*. Various translations.

Vallée, J. (1969) *Passport to Magonia: From Folklore to Flying Saucers*. Chicago, IL: Henry Regnery Company.

Keel, J.A. (1975) *The Eighth Tower*. New York: Saturday Review Press.

Chen, F.F. (2016) *Introduction to Plasma Physics and Controlled Fusion*. 3rd edition. Cham: Springer.

Fortov, V.E., Ivlev, A.V., Khrapak, S.A., Khrapak, A.G. and Morfill, G.E. (2005) *Dusty Plasmas*. Physics-Uspekhi, 47(5), pp. 447–492. Moscow: Russian Academy of Sciences.

18 Sky Trumpets as Atmospheric Voices

Historical and Primary Sources

Pliny the Elder (1938) *Natural History, Volume I: Books 1–2*. Translated by H. Rackham. Loeb Classical Library. Cambridge, MA: Harvard University Press.

Legge, J. (trans.) (1865) *The Chinese Classics, Volume III: The Shoo King (Shujing)*. London: Trübner & Co.

Durrant, S., Li, W. and Schaberg, D. (2016) *Zuo Tradition / Zuozhuan*. Seattle: University of Washington Press.

Paracelsus (1658) *A Book on Nymphs, Sylphs, Pygmies, and Salamanders, and on the Other Spirits*. Basel.

Atmospheric Acoustics and Infrasound

Le Pichon, A., Blanc, E. and Hauchecorne, A. (2005) *Infrasound Monitoring for Atmospheric Studies*. Dordrecht: Springer.

Campus, P. and Christie, D.R. (2010) 'Worldwide observations of infrasonic waves', *Infrasound Monitoring for Atmospheric Studies*, pp. 185–234.

Bedard, A.J. and Georges, T.M. (2000) 'Atmospheric infrasound', *Physics Today*, 53(3), pp. 32–37.

Evers, L.G. and Haak, H.W. (2010) 'The characteristics of infrasound, its propagation and some early history', *Journal of Atmospheric and Solar-Terrestrial Physics*, 72(10), pp. 1159–1165.

Sky Trumpet Case Studies and Modern Reports

MacLeod, D. (2012) 'The "Skyquakes" and "Sky Trumpets" mystery', *Journal of Scientific Exploration*, 26(3), pp. 417–433.

Keay, C.S.L. (1980) 'Anomalous sounds associated with earthquakes', *Bulletin of the Seismological Society of America*, 70(6), pp. 2061–2074.

Kiev (2011), Slovakia (2016), British Columbia (2020).

Bioacoustics and Cetacean Analogues

Payne, R. and McVay, S. (1971) 'Songs of humpback whales', *Science*, 173(3997), pp. 585–597.

Au, W.W.L. and Hastings, M.C. (2008) *Principles of Marine Bioacoustics*. New York: Springer.

Tyack, P.L. (2008) 'Implications for marine mammals of large-scale changes in the marine acoustic environment', *Journal of Mammalogy*, 89(3), pp. 549–558.

Resonance, Cavities, and Acoustic Scaling

Fletcher, N.H. and Rossing, T.D. (1998) *The Physics of Musical Instruments*. 2nd edition. New York: Springer.

Morse, P.M. and Ingard, K.U. (1986) *Theoretical Acoustics*. Princeton, NJ: Princeton University Press.

Rossing, T.D. (2007) *Springer Handbook of Acoustics*. New York: Springer.

Plasma Physics and Self-Organisation

Chen, F.F. (2016) *Introduction to Plasma Physics and Controlled Fusion*. 3rd edition. Cham: Springer.

Shukla, P.K. and Mamun, A.A. (2002) *Introduction to Dusty Plasma Physics*. Bristol: Institute of Physics Publishing.

Nicolis, G. and Prigogine, I. (1977) *Self-Organization in Nonequilibrium Systems*. New York: Wiley.

19 Crop Circles

Atmospheric Plasma, Vortices, and Electromagnetic Phenomena

Andreopoulos, Y. & Xanthos, S. (1998). *Aspects of Vortex Dynamics and Turbulence in Fluids.* Applied Mechanics Reviews, 51(5), 371–390.

Ashkenazy, Y., Havlin, S., & Dinar, N. (1999). 'Electromagnetic signatures of atmospheric vortices'. *Physical Review Letters*, 82(16), 3152–3155.

Draine, B. (2011). *Physics of the Interstellar and Intergalactic Medium.* Princeton University Press.

Piel, A. (2017). *Plasma Physics: An Introduction to Laboratory, Space, and Fusion Plasmas.* Springer.

Stenhoff, M. (1999). *Ball Lightning: An Unsolved Problem in Atmospheric Physics.* Springer.

Plant Physiology, Microwave Heating, and Node Expansion

Burton, N. & Oliver, R. (1989). 'Microwave-induced elongation in plant stem nodes.' *Journal of Experimental Botany*, 40(3), 345–352.

Hagström, O. & Olsson, G. (1991). 'Dielectric heating effects in biological tissues at microwave frequencies.' *Bioelectromagnetics*, 12(3), 147–160.

Levengood, W.C. (1994). 'Anatomical and biophysical changes in plants from crop formations.' *Physiologia Plantarum*, 92(2), 356–363.

Levengood, W.C. & Talbott, N.P. (1999). 'Dispersion of energy in plant tissues exposed to rapid heating mechanisms.' *Journal of Scientific Exploration*, 13(4), 571–595.

Behaviour, Ethology, Emergent Pattern Formation

Ballerini, M. et al. (2008). 'Interaction ruling animal collective behaviour depends on topological rather than metric distance.' *Proceedings of the National Academy of Sciences*, 105(4), 1232–1237.

Couzin, I.D. (2009). 'Collective cognition in animal groups.' *Trends in Cognitive Sciences*, 13(1), 36–43.

Parrish, J.K. & Edelstein-Keshet, L. (1999). 'Complexity, pattern, and evolutionary trade-offs in animal aggregation.' *Science*, 284(5411), 99–101.

Zhang, H. et al. (2010). *Self-organised dynamics in fluids and biological systems.* Nature Physics, 6, 364–370.

Electromagnetic Interaction With Biological Material

Foster, K.R. & Cheever, E.A. (1992). 'Microwave heating of biological tissues.' *IEEE Transactions on Biomedical Engineering*, 39(9), 915–927.

Mittra, R. & Itoh, T. (2000). *Electromagnetic Waves and Antennas.* IEEE Press.

General Documentation of Biophysical Anomalies

Haselhoff, E.H. (2001). *The Deepening Complexity of Crop Circles: Scientific Evidence.* Frog Books.

Randles, J. (1999). *Crop Circles: A Mystery Solved?* Blandford.

Silva, L. & Levengood, W.C. (1996). 'Biophysical evidence for unusual energy deposition in crop formations.' *Physiologia Plantarum*, 98(2), 390–396.

Plasma Behaviour and Self-Organisation - General

Bellan, P. (2006). *Fundamentals of Plasma Physics.* Cambridge University Press.

Helliwell, R. (1965). *Whistlers and Related Ionospheric Phenomena.* Stanford University Press.

(Relevant for structured plasma waves in the atmosphere.)

Shukla, P.K. & Eliasson, B. (2009). 'Colloquium: Nonlinear collective interactions in plasma systems.' *Reviews of Modern Physics*, 81(1), 25–40.

20 Dragons and Sky Serpents

Birkeland currents and filamentary plasma

Alfvén, H. (1986) 'Double layers and circuits in astrophysics', *IEEE Transactions on Plasma Science*, 14(6), pp. 779–793.

Peratt, A.L. (1992) *Physics of the Plasma Universe.* New York: Springer.

McPherron, R.L. (1995) 'Magnetospheric dynamics', in Kivelson, M.G. and Russell, C.T. (eds) *Introduction to Space Physics.* Cambridge: Cambridge University Press, pp. 400–458.

Sprites, jets, ELVES and large-scale atmospheric plasma

Pasko, V.P. (2010) 'Recent advances in theory of transient luminous events', *Journal of Geophysical Research: Space Physics*, 115(A00E35), pp. 1–24.

Neubert, T. et al. (2008) 'Recent results from studies of electric discharges in the mesosphere', *Surveys in Geophysics*, 29(2), pp. 71–137.

Lyons, W.A. (1996) 'Sprite observations above the U.S. High Plains in relation to their parent thunderstorm systems', *Journal of Geophysical Research: Atmospheres*, 101(D23), pp. 29641–29652.

Ball lightning and luminous "fire-like" plasma

Stenhoff, M. (1999) *Ball Lightning: An Unsolved Problem in Atmospheric Physics.* New York: Springer.

Barry, J.D. (1980) *Ball Lightning and Bead Lightning.* New York: Plenum Press.

Dusty plasma and self-organisation

Morfill, G.E. and Ivlev, A.V. (2009) 'Complex plasmas: An interdisciplinary research field', *Reviews of Modern Physics*, 81(4), pp. 1353–1404.

Fortov, V.E. and Morfill, G.E. (eds) (2010) *Complex and Dusty Plasmas: From Laboratory to Space.* Boca Raton: CRC Press.

Myth, dragons, and cross-cultural serpent traditions

Dragons and serpents in comparative myth

Fontenrose, J. (1959) *Python: A Study of Delphic Myth and Its Origins.* Berkeley: University of California Press.

Loomis, R.S. (1948) *Celtic Myth and Arthurian Romance.* New York: Columbia University Press.

Littleton, C.S. and Malcor, L.A. (2000) *From Scythia to Camelot: A Radical Reassessment of the Legends of King Arthur, the Knights of the Round Table, and the Holy Grail.* New York: Routledge.

Barber, E.W. and Barber, P.T. (2004) *When They Severed Earth from Sky: How the Human Mind Shapes Myth.* Princeton, NJ: Princeton University Press.

Chinese & Japanese dragon/serpent traditions

Eberhard, W. (1968) *The Local Cultures of South and East China.* Leiden: Brill.

Birrell, A. (1993) *Chinese Mythology: An Introduction.* Baltimore, MD: Johns Hopkins University Press.

Mesoamerican dragon/serpent traditions

Miller, M.E. and Taube, K. (1993) *An Illustrated Dictionary of the Gods and Symbols of Ancient Mexico and the Maya.* London: Thames & Hudson.

Australian Aboriginal Rainbow Serpent

Taçon, P.S.C., Fullagar, R., Ouzman, S. and Mulvaney, K. (1996) 'Cupules, "cupule rock shelters" and the earliest

Australian rock engravings', *Proceedings of the Prehistoric Society*, 62, pp. 1–30.

Radin, D. and Rainbird, P. (2003) 'The Rainbow Serpent Myth of Arnhem Land', *The Artefact*, 26, pp. 33–47.

Shamanic serpents, Ayahuasca imagery & Altered States

Narby, J. (1998) *The Cosmic Serpent: DNA and the Origins of Knowledge*. New York: Tarcher/Putnam.

Shanon, B. (2002) *The Antipodes of the Mind: Charting the Phenomenology of the Ayahuasca Experience*. Oxford: Oxford University Press.

Lewis-Williams, D. (2002) *The Mind in the Cave: Consciousness and the Origins of Art*. London: Thames & Hudson.

Winkelman, M. (2010) *Shamanism: A Biopsychosocial Paradigm of Consciousness and Healing*. 2nd edition. Santa Barbara, CA: Praeger.

Vision, gating, and brain–field interaction

Ward, J. (2019) *The Student's Guide to Cognitive Neuroscience*. 4th edition. London: Routledge.

Llinás, R. (2001) *I of the Vortex: From Neurons to Self*. Cambridge, MA: MIT Press.

Friston, K. (2010) 'The free-energy principle: A unified brain theory?', *Nature Reviews Neuroscience*, 11(2), pp. 127–138.

Persinger, M.A. (1983) 'Religious and mystical experiences as artifacts of temporal lobe function: A general hypothesis', *Perceptual and Motor Skills*, 57(3), pp. 1255–1262.

Wolfram, Ruliad and computational slices

Wolfram, S. (2020) *A Project to Find the Fundamental Theory of Physics*. Champaign, IL: Wolfram Media.

Wolfram, S. (2021) 'The Ruliad: A Multi-Computational Definition of the Universe', *Complexity*, 2021, Article ID 5573610.

21 Disruption, Humanity and Contact

All-domain Anomaly Resolution Office (AARO) (n.d.) *AARO Home: Unidentified Anomalous Phenomena (UAP) Definitions and Reporting.* Washington, DC: U.S. Department of Defense.

Office of the Director of National Intelligence (2024) *2024 Consolidated Annual Report on Unidentified Anomalous Phenomena.* Washington, DC: ODNI.

United States Congress (n.d.) *Unidentified Anomalous Phenomena Reporting Procedures (50 U.S.C. § 3373b).* Washington, DC: Office of the Law Revision Counsel, U.S. House of Representatives.

NASA (2022) *Unidentified Anomalous Phenomena (UAP): NASA Study.* Washington, DC: National Aeronautics and Space Administration.

United States Department of Defense (2024) *Report on Unidentified Anomalous Phenomena.* Office of the Director of National Intelligence. Washington, DC: U.S. Government Publishing Office.

United States Congress (2025) *House Permanent Select Committee on Intelligence: Unidentified Anomalous Phenomena Oversight Hearing.* Washington, DC: U.S. Congress.

United States Air Force (2024) *MQ-9 Reaper Remotely Piloted Aircraft Fact Sheet.* Washington, DC: Department of the Air Force.

United States Department of Defense (2024) *AGM-114 Hellfire Missile Systems Overview.* Washington, DC: U.S. Department of Defense.

Office of the Director of National Intelligence (2025) *Annual Report on Unidentified Anomalous Phenomena.* Washington, DC: ODNI.

Hastings, R. (2017) *UFOs and Nukes: Extraordinary Encounters at Nuclear Weapons Sites.* 2nd edition. Charleston, SC: History Press.

Salas, R.L. (2010) *Faded Giant: The UFO Mystery at Malmstrom Air Force Base.* Raleigh, NC: Lulu Press.

Vallée, J. (1969) *Passport to Magonia: From Folklore to Flying Saucers.* Chicago, IL: Henry Regnery Company.

Keel, J.A. (1975) *The Eighth Tower.* New York: Saturday Review Press.

Strand, E., Faugstad, Ø., & Strand, G. (2014) *The Hessdalen Phenomenon: Thirty Years of Scientific Investigation.* Østfold, Norway: Østfold University College.

Teodorani, M. (2004) *A Long-Term Scientific Survey of the Hessdalen Phenomenon.* Modena: Italian National Research Council.

Teodorani, M. (2014) *Unidentified Flying Objects: Observations, Explanations and Speculations.* Berlin: Springer.

Stenhoff, M. (1999) *Ball Lightning: An Unsolved Problem in Atmospheric Physics.* New York: Springer.

Singer, S.J. (1971) *The Nature of Ball Lightning.* New York: Plenum Press.

Chen, F.F. (2016) *Introduction to Plasma Physics and Controlled Fusion.* 3rd edition. Cham: Springer.

Fortov, V.E., Ivlev, A.V., Khrapak, S.A., Khrapak, A.G. and Morfill, G.E. (2005) *Dusty Plasmas.* Physics-Uspekhi, 47(5), pp. 447–492. Moscow: Russian Academy of Sciences.

Smirnov, B.M. (1993) *Physics of Ionized Gases.* New York: Wiley.

Abraham, R. & Shaw, C. (1992). *Dynamics: The Geometry of Behaviour.* Addison-Wesley, Redwood City.

Akers, B. (2017). 'Electromagnetic field effects on the human nervous system: Evidence and plausibility', *Journal of Consciousness Studies*, 24(9–10), pp. 74–103.

Aveni, A. (1993). *The Book of the Sky: Ancient Cosmologies and Celestial Myths.* Thames & Hudson, London.

Barry, R. J. (2015). 'EEG coherence, synchronisation, and the sensitivity of cortical networks to external fields', *Clinical Neurophysiology*, 126(11), pp. 2063–2074.

Beers, B. (2004). *Lilith and Her Sisters: Female Demons in Mesopotamian Tradition.* Brill, Leiden.

Bekker, C. (2019). 'Sleep paralysis and the neurocognitive boundary state', *Sleep Medicine Reviews*, 46, pp. 3–11.

Bender, J. & Wellmann, J. (eds.) (2018). *Spirits in the Material World: Cognitive and Cultural Perspectives on Apparitions.* University of Chicago Press, Chicago.

Blyth, C. (2020). *A Feminist Companion to the Bible: Lilith, Night and Demonology in Ancient Near Eastern Texts.* Oxford University Press, Oxford.

Caldwell, R. (2018). 'Aboriginal Min Min lights: An anthropological overview', *Oceania*, 88(1), pp. 30–52.

Carrington, P. (2017). *Shamanic States and Altered Consciousness.* Routledge, London.

Coppens, P. (2002). *The Watchers: The Hidden World of Ancient Angels.* Random House, New York.

Davis, M. (2005). 'Electromagnetic sensitivity of the limbic system: implications for anomalous experiences', *Perceptual and Motor Skills*, 100(3), pp. 851–866.

Devereux, P. (1996). *Earth Lights: The Visible Evidence.* Blandford, London.

Dunn, F. & Marsh, S. (2020). 'The interaction of low-frequency electromagnetic fields with cortical rhythms', *Neuroscience Letters*, 731, pp. 135023.

Goodman, F. (1990). *How About Demons? Possession and Exorcism in the Modern World.* Indiana University Press, Bloomington.

Hamacher, D. W. & Norris, R. P. (2011). 'Bridging Aboriginal astronomy and Western astrophysics: Min Min lights', *Australian Journal of Anthropology*, 22(2), pp. 175–194.

Hijzen, C. & Markus, T. (2019). 'Sleep paralysis: hallucination, folklore, or anomalous experience?', *Consciousness and Cognition*, 71, pp. 1–13.

Hitchcock, H. (1994). *The Entity: A True Story of a Haunting.* Jove Books, New York.

Jung, C. G. (1960). *Synchronicity: An Acausal Connecting Principle.* Princeton University Press, Princeton.

Kaku, M. (2016). *Physics of the Impossible.* Penguin Books, London.

Lavoie, J. (2021). 'Electromagnetic anomalies in UAP encounters: A review of case reports', *Journal of Scientific Exploration*, 35(4), pp. 739–762.

Lewis, J. (2003). *Demons and Spirits in Ancient Judaism.* Cambridge University Press, Cambridge.

McFadden, J. (2020). 'Integrating information in the brain: electrical fields and consciousness', *Neuroscience of Consciousness*, 2020(1), niaa016.

Persinger, M. A. (1987). *Neuropsychological Bases of God Beliefs.* Praeger, New York.

Persinger, M. A. (2014). 'The sensed presence and electromagnetic fields: Clinical and experimental evidence', *Psychological Reports*, 115(2), pp. 605–626.

Persinger, M. A. & Healey, F. (2002). 'Experimental induction of the "sensed presence" using varying electromagnetic fields', *Perceptual and Motor Skills*, 94(3), pp. 1257–1268.

Pettazzoni, R. (1954). *The All-Knowing God: Religious Belief Systems of Antiquity*. Brill, Leiden.

Pincombe, B. (2019). *Dreamtime and the Spirit World: Indigenous Cosmology and Experience*. Allen & Unwin, Sydney.

Rogers, L. & Wulff, D. (2021). 'Electromagnetic effects in poltergeist and haunting cases: An integrative review', *Journal of Parapsychology*, 85(2), pp. 220–245.

Slater, T. & Lewis, P. (2018). 'Field coupling and the biological brain: susceptibility and thresholds', *Frontiers in Human Neuroscience*, 12, article 87.

Sturrock, P. (1999). *The UFO Enigma: A New Review of the Physical Evidence*. Warner Books, New York.

Taff, B. (2014). *Aliens Above, Ghosts Below*. Cosmic Pantheon Press, Los Angeles.

Tiller, W. (1997). *Science and Human Transformation: Subtle Energies and Intentionality*. Pavior Publishing, Walnut Creek.

Vernon, D. (2005). 'Theta rhythms and altered states: implications for external field influence', *Consciousness and Cognition*, 14(3), pp. 575–587.

Young, S. & Goulet, J. (2016). *Being Changed: Indigenous Encounters with Spirits and Non-Human Beings*. University of Toronto Press, Toronto.

23 Storms as Sentient Structure

Atmospheric Electricity & Lightning Physics

Rakov, V. A., & Uman, M. A. (2003). *Lightning: Physics and Effects.* Cambridge University Press.

Dwyer, J. R., & Uman, M. A. (2014). *The physics of lightning. Physics Reports*, 534(4), 147–241.

Williams, E. R. (2005). *Lightning and climate: A review. Atmospheric Research*, 76(1–4), 272–287.

Plasma Physics & Electromagnetic Confinement

Chen, F. F. (2016). *Introduction to Plasma Physics and Controlled Fusion.* Springer.

Alfvén, H. (1981). *Cosmic Plasma.* Springer.

Haines, M. G. (2011). *A review of the pinch effect. Plasma Physics and Controlled Fusion*, 53(9).

Ball Lightning

Stenhoff, M. (1999). *Ball Lightning: An Unsolved Problem in Atmospheric Physics.* Springer.

Jennison, R. C. (1969). *Ball lightning. Nature*, 224, 895–896. *(Primary source describing the 1953 aircraft encounter.)*

Cen, J., Yuan, P., & Xue, S. (2014). *Observation of the optical and spectral characteristics of ball lightning. Physical Review Letters*, 112(3).

Plasma Structures, Geometry & Resonance

Shukla, P. K., & Mamun, A. A. (2002). *Introduction to Dusty Plasma Physics.* Institute of Physics Publishing.

Tsytovich, V. N., et al. (2007). *Self-organization phenomena in complex plasmas. New Journal of Physics*, 9.

Bostick, W. H. (1956). *Experimental study of ionised matter projected across a magnetic field. Physical Review*, 104(2).

Matter, Electromagnetism & "Solidity"

Feynman, R. P., Leighton, R. B., & Sands, M. (1964). *The Feynman Lectures on Physics, Vol. II.* Addison-Wesley.

Purcell, E. M., & Morin, D. J. (2013). *Electricity and Magnetism.* Cambridge University Press.

Biology, Life Criteria & Metaphor

Cleland, C. E., & Chyba, C. F. (2002). *Defining 'life'. Origins of Life and Evolution of the Biosphere*, 32, 387–393.

Forterre, P. (2010). *Defining life: The virus viewpoint. Origins of Life and Evolution of Biospheres*, 40, 151–160.

Mythology, Anthropology & Interpretation

Eliade, M. (1964). *Shamanism: Archaic Techniques of Ecstasy.* Princeton University Press.

Frazer, J. G. (1922). *The Golden Bough.* Macmillan.

Historical Scientific Resistance & Delayed Validation

Oreskes, N. (1999). *The rejection of continental drift. Oxford University Press.*

Csillaghy, A., et al. (2018). *Imaging observations of the Kordylewski dust clouds. Monthly Notices of the Royal Astronomical Society*, 475(1).

24 Conclusion

Philosophy of Science and Paradigm Change

Kuhn, T. S. (1962). *The Structure of Scientific Revolutions.* Chicago: University of Chicago Press.

Popper, K. R. (1959). *The Logic of Scientific Discovery.* London: Routledge.

Lakatos, I. (1978). *The Methodology of Scientific Research Programmes.* Cambridge: Cambridge University Press.

Complexity, Emergence, and Non-Biological Organisation

Mitchell, M. (2009). *Complexity: A Guided Tour.* Oxford: Oxford University Press.

Kauffman, S. A. (1995). *At Home in the Universe: The Search for Laws of Self-Organization and Complexity*. Oxford: Oxford University Press.

Perception, Observation and Instrumentation Limits

Hecht, E. (2017). *Optics* (5th ed.). Harlow: Pearson.

Land, E. H. (1977). The retinex theory of colour vision. *Scientific American*, 237(6), 108–128.

Historical Cases of Initially Rejected Science

Bretz, J. H. (1923). The channelled scablands of the Columbia Plateau. *Journal of Geology*, 31(8), 617–649.

Wegener, A. (1915). *Die Entstehung der Kontinente und Ozeane*. Braunschweig: Vieweg.

Kordylewski, K. (1961). Dust clouds around the Earth–Moon libration points. *Acta Astronomica*, 11, 165–171.

Slíz-Balogh, J., et al. (2018). Imaging the Kordylewski dust cloud. *Monthly Notices of the Royal Astronomical Society*, 480(4), 5570–5578.

Made in the USA
Monee, IL
05 November 2026

67815695R00216